Elements in the Philosophy of Mathematics
edited by
Penelope Rush
University of Tasmania
Stewart Shapiro
The Ohio State University

THE EUCLIDEAN PROGRAMME

A. C. Paseau
University of Oxford
Wesley Wrigley
London School of Economics and Political Science

Shaftesbury Road, Cambridge CB2 8EA, United Kingdom

One Liberty Plaza, 20th Floor, New York, NY 10006, USA

477 Williamstown Road, Port Melbourne, VIC 3207, Australia

314–321, 3rd Floor, Plot 3, Splendor Forum, Jasola District Centre,
New Delhi – 110025, India

103 Penang Road, #05–06/07, Visioncrest Commercial, Singapore 238467

Cambridge University Press is part of Cambridge University Press & Assessment,
a department of the University of Cambridge.

We share the University's mission to contribute to society through the pursuit of
education, learning and research at the highest international levels of excellence.

www.cambridge.org
Information on this title: www.cambridge.org/9781009494403

DOI: 10.1017/9781009221955

First published 2024

A catalogue record for this publication is available from the British Library

ISBN 978-1-009-49440-3 Hardback
ISBN 978-1-009-22198-6 Paperback
ISSN 2399-2883 (online)
ISSN 2514-3808 (print)

The Euclidean Programme

Elements in the Philosophy of Mathematics

DOI: 10.1017/9781009221955
First published online: January 2024

A. C. Paseau
University of Oxford

Wesley Wrigley
London School of Economics and Political Science

Author for correspondence: A. C. Paseau,
alexander.paseau@philosophy.ox.ac.uk

Abstract: The Euclidean Programme embodies a traditional sort of epistemological foundationalism, according to which knowledge – especially mathematical knowledge – is obtained by deduction from self-evident axioms or first principles. Epistemologists have examined foundationalism extensively, but neglected its historically dominant Euclidean form. By contrast, this book offers a detailed examination of Euclidean foundationalism, which, following Lakatos, the authors call the Euclidean Programme. The book rationally reconstructs the programme's key principles, showing it to be an epistemological interpretation of the axiomatic method. It then compares the reconstructed programme with select historical sources: Euclid's *Elements*, Aristotle's *Posterior Analytics*, Descartes's *Discourse on Method*, Pascal's *On the Geometric Mind* and a twentieth-century account of axiomatisation. The second half of the book philosophically assesses the programme, exploring whether various areas of contemporary mathematics conform to it. The book concludes by outlining a replacement for the Euclidean Programme.

Keywords: Euclideanism, epistemological foundationalism, Pascal, mathematical knowledge, axioms

ISBNs: 9781009494403 (HB), 9781009221986 (PB), 9781009221955 (OC)
ISSNs: 2399-2883 (online), 2514-3808 (print)

Contents

1 Introduction

In a short philosophical piece penned in the 1930s, Einstein writes rapturously about the beginnings of Western science:

> We honour ancient Greece as the cradle of western science. She for the first time created the intellectual miracle of a logical system, the assertions of which followed one from another with such rigor that not one of the demonstrated propositions admitted of the slightest doubt. (1934: 164)

Einstein then names the miraculous logical system he has in mind and adds by way of comment:

> This marvellous accomplishment of reason gave to the human spirit the confidence it needed for its future achievements. (1934: 164)

The system Einstein had in mind, you might have guessed, is that of Euclid's geometry in the *Elements* (c. 300 BC). Einstein is a recent figure in a long line of those who have admired the *Elements* as a paragon of mathematical method. Euclid's text took pride of place in at least three brilliant mathematical cultures – ancient Greek, mediaeval Arabic, and early modern European – and was a cornerstone of the school curriculum in the West from the Renaissance until the twentieth century. Hailed as a shining example of the mathematical method, in fact of method *tout court*, the *Elements* spawned hundreds of imitators, not just in geometry but in many other fields too.

So, what is the method of Euclid's *Elements*? Starting from some definitions, postulates, and common notions, Euclid derives the geometry of his day theorem by theorem, in a cumulative manner over the course of thirteen books. Book I's postulates and common notions are as follows:[1]

Postulates

Let the following be postulated:

1. To draw a straight line from any point to any point.
2. To produce a finite straight line continuously in a straight line.
3. To describe a circle with any centre and distance.
4. That all right angles are equal to one another.
5. That, if a straight line falling on two straight lines makes the interior angles on the same side less than two right angles, the two straight lines, if produced indefinitely, meet on that side on which are the angles less than the two right angles.

[1] For Euclid's text, we have used Heath (1925).

Common Notions

1. Things which are equal to the same thing are also equal to one another.
2. If equals be added to equals, the wholes are equals.
3. If equals be subtracted from equals, the remainders are equal.
4. Things which coincide with one another are equal to one another.
5. The whole is greater than the part.

Euclid supplements these with twenty-three definitions in Book I (omitted here), including further facts about angles and triangles.

Given the importance of the Euclidean method to the epistemology of mathematics and other fields, it is surprising to find so little attention devoted to it in recent philosophy. Imre Lakatos is one of the very few philosophers of the recent past to have written about what he calls 'the Euclidean Programme'. We shall use Lakatos's characterisation as a springboard for ours and adopt his name, abbreviating 'Euclidean Programme' as 'EP'. Epistemologists have, of course, examined foundationalism more generally, but they have neglected its more specific, and historically dominant, instance: the EP as it has been conceived over the centuries. Against this trend, the present essay is devoted to examining the EP.

First of all, we must clarify that the EP is not to be conflated with the axiomatic method in mathematics. The axiomatic method in general is of huge importance, mathematically, historically, and philosophically. And, of course, Euclid's *Elements* both pioneers the method and is a paradigm of it. But the Euclidean Programme is a particular philosophical take on the axiomatic method and goes beyond mere practice of the method. It will be the focus of our attention here.[2]

As to what the EP actually *is*, we propose a rational reconstruction of its key principles in §2. This reconstruction tries to model what people who have been inspired by the *Elements* have maintained. Like any such reconstruction, ours does not correspond to a historically attested expression; rather, it draws together some key ideas behind various expressions. Although we are more interested in philosophical analysis of the EP than in its long history, a historical overview will nevertheless be useful. We take the apogee of the EP to be in the early modern period, specifically the seventeenth century. We compare our reconstruction of the EP with three historical accounts: Aristotle's discussion of scientific method in the *Posterior Analytics* (§4), which predates Euclid, and two seventeenth-century versions, in Descartes' *Discourse on Method* and Pascal's *On the Geometric Mind* respectively (both in §5). Before that, we say a few words about the *Elements* (§3),

[2] We will not, therefore, be discussing some of the most important figures in axiomatic mathematics (e.g. David Hilbert), or some of its most important features, such as the organisation of a mathematical subfield, in significant detail.

the main point being to caution the reader against confusing the EP with how Euclid actually proceeds in the *Elements*. We conclude the more historical discussion with a schematic twentieth-century account of Descriptive Axiomatisation (§6). History is then followed by philosophical assessment. §§7–8 critically assess the EP, and §9 sketches what should replace it. §10 very briefly concludes.

All in all, the present essay offers a combined historical and critical analysis of the Euclidean Programme.[3] We try to impose some structure on a historical jumble of ideas (§§3–6), but we also argue for a position about the EP's current status (§§7–9).

2 The Euclidean Programme

The term 'Euclidean Programme' is borrowed from Lakatos, whose paper prompted our interest in delineating it. Lakatos contrasts the Euclidean Programme with an Empiricist one:

> The Euclidean programme proposes to build up Euclidean theories with foundations in meaning and truth-value at the top, lit by the *natural light of Reason*, specifically by arithmetical, geometrical, metaphysical, moral, etc. intuition. The Empiricist programme proposes to build up Empiricist theories with foundations in meaning and truth-value at the bottom, lit by the *natural light of Experience*. Both programmes however rely on Reason (specifically on logical intuition) for the safe transmission of meaning and truth-value. (Lakatos 1962: 5)

We return to the Empiricist Programme in §9 and until then concentrate on the Euclidean one. Lakatos describes the latter in more detail in the following passage:

> I call a deductive system a 'Euclidean theory' if the propositions at the top (*axioms*) consist of perfectly well-known terms (*primitive terms*), and if there are *infallible truth-value-injections* at this top of the truth-value *True*, which flows downwards through the deductive channels of truth-transmission (*proofs*) and inundates the whole system. (If the truth-value at the top was *False*, there would of course be no current of truth-value in the system.) Since the Euclidean programme implies that all knowledge can be deduced from a finite set of trivially true propositions consisting only of terms with a trivial meaning-load, I shall call it also the *Programme of Trivialization of Knowledge*. Since a Euclidean theory contains only indubitably true propositions, it operates neither with conjectures nor with refutations. In a fully-fledged Euclidean theory meaning, like truth, is injected at the top and it flows down safely through meaning-preserving channels of nominal definitions from the primitive terms to the (abbreviatory and therefore theoretically superfluous) defined terms. A Euclidean theory is *eo ipso* consistent, for all the propositions occurring in it are true, and a set of true propositions is certainly consistent. (Lakatos 1962: 4–5)

[3] Other authors have used different names for what, following Lakatos, we call the Euclidean Programme. For example, Shapiro (2009: 181) calls it *Euclidean Foundationalism*.

In this passage, Lakatos speaks of truth and meaning injection, but this is somewhat misleading. The EP represents an epistemological conception, and the hierarchical path from axioms to theorems is a path followed by a subject.[4] The flow metaphor is better construed as the transmission of an epistemic good of some sort, such as justification. The picture is then a foundationalist one in which one gains justification for the axioms first and thence for theorems by inferring these from the axioms.

Lakatos also calls the axioms 'trivially true' and says they bear a 'trivial meaning-load'. We don't know what exactly Lakatos meant by the word 'trivial'. One way to understand it is as the broadly empiricist idea, favoured by Hume and the logical empiricists: mathematical statements are true in virtue of meaning and therefore empty of content.[5] If so, we part ways with Lakatos: it is entirely compatible with the EP that axioms are substantive. For example, recognition of the axioms' truth could be the product of mathematical intuition, a faculty distinct from any that informs us of the trivial truth of statements such as 'bachelors are unmarried'. We take the idea at the heart of the EP to be that axioms are self-evident, and we remain neutral on their 'triviality' (whatever exactly this means).

Lakatos then goes on to make a very acute observation – the key, we believe, to understanding the Euclidean Programme:

> We can get a long way merely by discussing *how* anything flows in a deductive system without discussing the problem of *what in fact flows* there, infallible truth or only, say, Russellian 'psychologically incorrigible' truth, Braithwaitian 'logically incorrigible' truth, Wittgensteinian 'linguistically incorrigible' truth or Popperian corrigible falsity and 'verisimilitude', Carnapian probability. (1962: 6)

Earlier, we spoke of an epistemic good flowing from axioms to theorems. This is the right way to characterise the EP if we are to maintain generality and avoid, or at least minimise, anachronism. The insight we extract from Lakatos is that we can achieve this by considering *how* the epistemic good flows rather than *what* it is. Succinctly, the EP is all about *Euclidean hydraulics*. An analogy: think of the Phillips machine, a post-war hydraulic model of the economy. Its inventor, Bill Philips, used it to demonstrate how money moves through an economy by letting coloured water flow through clear pipes.[6] In our epistemic analogue, the coloured water corresponds to the epistemic good. It is injected at the top, where the axioms lie, and thence flows down to the theorems.

[4] It is unclear what it might literally mean for truth and meaning to flow down some channel.

[5] If axioms were so trivial as to be logical then they would be unnecessary, as they would be delivered by the logic. But we take it Lakatos has a broader sense of triviality in mind.

[6] Readers should google 'MONIAC' for a demonstration of the machine at work.

Different versions of the EP will differ on what exactly the good is. We note, however, that not every epistemic good possessed by the axioms will flow down the relevant channels; in particular the axioms' *self-evidence* (about which more shortly) may not be transferred to the theorems by inference.

Another important point is that our understanding of a theory is broader than that of the contemporary logician, who understands it, roughly, as a deductively closed set of formal sentences in a formal logic. As we will see, the language of a theory putatively instantiating the EP does not have to be formal; it could be Greek, English, or any other language. It should be no part of the EP that an axiomatisation be formal, as that would be unfaithful to its history. Indeed, as Jonathan Barnes points out, the idea of a formal language was alien to ancient deductive thought.[7]

Nor do inferences in this context have to be purely logical.[8] Kantians might, for example, maintain that mathematical reasoning employs ineliminably mathematical modes of inference (say, spatial intuition in geometry); if so, the conclusion is in a strong, but not strictly logical, sense implied by the premises. So as not to restrict the EP's range of application too narrowly, we allow inferences that track these implications as part of the Euclidean picture. Moreover, which inferences one considers logical will be sensitive to the background logic, and the EP does not prescribe a particular background logic to be used. Indeed, advocacy of the EP is perfectly consistent with some version of an anti-logical view, as espoused by Descartes, for example (see §5.1). In short: a theory for us is simply a collection of sentences about a subject matter, closed under a relation that need not be formal or even logical.

To further clarify this point, consider the Kneales' account of the geometric method in their classic text *The Development of Logic*. The Kneales single out three ingredients in the 'customary presentation of geometry as a deductive science' (1962: 3). First, 'certain propositions of the science must be taken as true without demonstration'; second, 'all the other propositions of the science must be derived from these' (1962: 3). The last ingredient is at once the most distinctive and the most controversial of the three:

> … the derivation must be made without any reliance on geometrical assertions other than those taken as primitive, i.e. it must be *formal* or independent of the special subject matter discussed in geometry … [thus] elaboration of a deductive system involves consideration of the relation of logical consequence or entailment. (1962: 3–4)

---

[7] '[N]either they [the Stoics] nor any other ancient logician ever considered inventing an artificial language for the use of logic' (Barnes 2005: 512).

[8] Like others, we distinguish inference (a movement in thought) from implication or consequence (a relation among propositions). In this Element, we are almost exclusively interested in the former.

Kneale and Kneale do not clarify whether the 'or' in '*formal* or independent of the special subject matter' is supposed to present two alternatives (the second condition being different) or just one (the second spelling out the first). Whatever they intended, the idea that the deductive logic in any axiomatic presentation of geometry must be formal should be resisted. Even if, as we believe, logic is formal, it should be no requirement of a Euclidean account of geometry that its logic be formal. (Euclid's logic itself was certainly not, though in §3 we shall draw a contrast between the flesh-and-blood Euclid and the ideal he imperfectly manifested.)

What *is* clear is that Kneale and Kneale insist on derivations being strictly logical. But it is equally clear that they mean to characterise any 'customary' axiomatic theory, including Euclid's. To stipulate that such an axiomatisation's rules *must* be strictly logical seems too stringent a requirement; it risks, for example, making the *Elements* not a 'customary' axiomatisation, if Euclid's system is not strictly logical because it appeals to geometric insight in various places (see §3). More generally, there is no strong historical precedent, prior to the late nineteenth century at any rate,[9] for thinking that the rules in a Euclidean axiomatisation may not be topic-specific. It is better, then, to characterise rules more neutrally and not decree that they be formal, or strictly logical.

Returning to Lakatos, we note that, for him, primitive terms of a theory must be *perfectly* well-known (again, in virtue of their meaning being somehow trivial). As we see it, however, the Euclidean Programme is primarily an epistemology of mathematical propositions, not terms. Given the axioms' pride of place in the EP, our understanding of the primitive terms must be sufficiently clear to enable the mathematician to understand, and hence see the truth of, the axioms. But there is little justification for Lakatos's assertion that the primitive terms of a Euclidean theory must be perfectly understood, for this is not required for the axioms to be self-evident. It can be completely evident, for example, that anybody taller than a tall person is tall, even to someone with a less-than-perfect understanding of the predicate 'is tall'. Or, for a mathematical example, it would have been completely evident to an eighteenth-century mathematician that the identity mapping on the reals was a function, even in the absence of a clear understanding of what real-valued functions, or even the reals, are. So, we require only that the axioms be graspable, in the sense of being possible to understand, and self-evident to a mathematician who has grasped the meanings of the primitive terms to an extent which allows them to understand the axioms, whether or not their grasp of the primitive terms is perfect.

[9] The slightly oblique reference here is to Frege, who believed that the rules were (as we would now put it) topic-neutral.

To continue the hydraulic metaphor, we can tolerate some impurity in the water, so long as it does not affect the flow.

With all that in mind, let's try to express the general picture slightly more precisely. At the picture's core is a three-place epistemic relation relating a subject S to a proposition p to a certain degree d: we might formalise this as $E(S, p, d)$. ('E' is nicely suggestive of both 'Euclidean' and 'epistemic'.) There is a suppressed time index here, which we usually ignore, since it won't affect the discussion much. We take p to be a proposition, but with some minor adjustments it could equally well be taken to be a belief, or even a fact. Relation E is a placeholder for a more specific epistemic relation, which different proponents of the EP will want to construe in different ways, say as some species of justification or warrant. Talk of the subject's having the relevant 'epistemic good' is then simply another way of saying that the subject stands in this relation E (to p and to degree d). For some p, the subject S may stand in relation E to p to the maximal degree – call this *max*. This, according to the EP, is the case for the axioms, so long as S clearly grasps them. (Different versions of the EP will have a different story to tell about what clearly grasping the axioms amounts to.) As an illustration, if we equate E with justified belief this becomes: S is maximally justified in believing any axiom a. (A notion which in turn can be made more precise in different ways, depending on the precise type of justification in question.) In a limiting case, which our characterisation allows for but does not focus on, justification could be all or nothing, so that there are just two degrees. Moreover, axioms must, of course, be true, as must be the sentences inferred from them.

Next, the EP contains a principle governing E-flow, or transmission of the relevant epistemic good E. In a strong version, the degree d is preserved in an inference from the conjunction of an inference's premises to its conclusion; in a weaker version, it is more or less preserved. A special case of the strong version is when the subject S is in the highest epistemic state with respect to the (finitely many) axioms' conjunction A and thus, according to the principle, potentially so with respect to the theorems. In that case, if $E(S, A, max)$ and p follows from A, then S can reason her way to p from A using the appropriate rules; and if she does so then $E(S, p, max)$ will hold. The weaker version of the flow principle is that in such a case if $E(S, A, d)$ then $E(S, p, d^*)$ obtains for some d^* not much lower than d. The weaker version of the flow principle is vague, and deliberately so. Being too precise about d-transmission here would be anachronistic and risk obscuring important common features between different historical expressions of the EP. Instantiated by justification, the weaker version of the principle says that our justification for theorems derived in this way is high; but it allows that this justification may not be maximally high, allowing for some erosion of justification in the course of inferring theorems (about which more in §7.3).

We assume throughout that the subject *S* has no other epistemic access to the conclusion than that provided by inferring it from the axioms. So we ignore for example the following sort of case: *S* infers *p* from some premises, *S* knows that *p* is Emmy Noether's favourite theorem and also knows that Noether was a highly reliable mathematician. *S* might then legitimately be more confident in *p* than in the conjunction of the inference's premises. More generally, we ignore testimonial and other sources of evidence, the better to focus on EP's epistemology of *proof*. Another complication we largely ignore (although see §7.3) are cases in which *S* reasons to the same *p* in different ways – via different proofs – which might result in *S* standing in relation *E* to *p* to a higher degree than if *S* just reasoned to *p* in only one such way.

We are now ready to present our rational reconstruction of the EP, which is made up of three core principles and four further ones. This simple device will permit a thoroughgoing comparison of diverse historical figures in the Euclidean tradition and facilitate a comparison of their actual methodology to this reconstructed ideal. Of course, any relation between the two is bound to be loose and inexact; a perfect fit is not to be expected. The historian of philosophy must be careful to avoid attributing claims to past philosophers in terms they would not acquiesce to. Our aim is to relate the EP, stated *in vacuo*, to real historical conceptions. Although the point of the exercise is to show that the EP does relate interestingly to various historical expressions of 'Euclideanism', we must be careful not to confuse an abstract prototype with historical expressions that suggest or approximate it in some interesting fashion. Having said that, something would be amiss with our rational reconstruction if it did *not* display important similarities with these historical expressions, especially the seventeenth-century ones.

Delaying the historical comparisons for now, we summarise the three core principles of the EP as follows:

EP-Truth	All axioms and theorems are true.
EP-Self-Evidence	All axioms are self-evident. That is to say, they are all graspable and if a subject clearly grasps an axiom then she bears relation *E* to it to the maximal degree.
EP-Flow	If a conclusion follows from some premises, and the subject clearly grasps this, and bears relation *E* to these premises to a high degree, she thereby bears relation *E* to the conclusion to the same, or a similarly high, degree.

To reiterate a key point, it is crucial that the relation *E* not be further specified, to make the EP an umbrella conception large enough to cover many and varied historical instances. Choosing a specific relation for *E* would rule out some paradigm examples of the EP and obscure deep commonalities. It is also crucial that 'clearly grasps' be understood in the usual sort of way rather than a loaded one that makes *EP-Self-Evidence* and *EP-Flow* come out true by definition. (Faced with apparent counterexamples, an unyielding Euclidean could insist: 'if you don't take the axioms to be self-evident then you haven't clearly grasped them', and similarly for *EP-Flow*.)

As for less central principles, we list these four[10]:

EP-Finite	The axioms are finitely many.
EP-General	All axioms are general propositions.
EP-Independence	Each axiom is independent of the others.
EP-Completeness	All truths of a certain kind can be inferred from the axioms.

We explain these four principles in turn and justify their inclusion in the EP.

EP-Finite (present in the quotation from Lakatos) is implicit in pretty much all axiomatisations prior to the twentieth century. Of course, prior to the rise of modern logic, there was no way for mathematicians to distinguish between first-order axiom schemes and single second-order axioms when formulating a principle such as mathematical induction (see §7.2). To avoid anachronism, then, we regard a theory as satisfying *EP-Finite* if it has a finite presentation. This may naturally be understood as saying that the number of non-schematic axioms plus the number of schemes is finite or, slightly differently, that the axioms and schemes can be described in a finitary way.[11] One could strengthen *EP-Finite* by adding the requirement that this number of axioms be small – it would be odd if the number of axioms and schemes ran into the hundreds – but we will not do so here.

When we are considering theories with a precise distinction between axioms and rules of inference, it is natural to strengthen *EP-Finite* to require additionally that the number of inference rules be finite. For, as is well-known, axioms and rules of inference are, to a certain extent, interchangeable. For instance, instead of having 'if φ then ψ' as an axiom, one could have the inference rule 'from φ, infer ψ'. So, a system with infinitely many rules of inference does not qualify as finite in the relevant sense.[12] It would, however, be anachronistic to formulate *EP-Finite* in terms of inference rules. As Barnes highlights (1993: 139), ancient logicians did not sharply distinguish between inference rules and

[10] What's the rationale for elevating three of the seven principles to the 'core'? If you deny any of those three, it seems to us that you are definitely not a Euclidean foundationalist. This is less clear for the other four principles.

[11] The difference between the two will not affect anything in this Element.

[12] One might also wish to exclude systems containing inference rules that use infinitely many premises, such as the omega-rule in arithmetic.

axioms, and even significantly later (e.g. in the early modern period), mathematical texts typically do not contain explicit lists of inference rules. In such cases, we require simply that the number of axioms be finite, in the sense described earlier.

EP-General is also very standard. What exactly generality comes to is hard to state,[13] but it is often easy to recognise. For example, Peano Arithmetic's axiom that distinct numbers have distinct successors is general, as are any of Euclid's common notions in Book I. (For Peano Arithmetic's axioms, see §7.2; for Euclid's common notions, see §1.) The first three of Euclid's five postulates are also recognisably general; they are about any lines, points and circles with given properties. But the fourth, which states that all right angles are equal to one another, mentions angles of a particular type (as does the fifth). However, a right angle is easily defined in more general terms by exploiting the fact that two right angles make up a line. Similarly, the axioms of Peano Arithmetic mention the number 0, yet 0 may be defined as the only natural number that satisfies the equation $x + x = x$. We may thus distinguish two ways in which an axiom may fail to be general. The first is by including terms for specific entities; the second is by including terms for entities that are not definable using general vocabulary.[14] It is only the latter that falls foul of *EP-General*, for any occurrence of an expression definable using general vocabulary is eliminable in favour of the definition.

EP-Independence requires of each axiom that there is no proof of it from the other axioms of the theory.[15] The concept of independence is most familiar from the history of the fifth postulate. This postulate, which we shall refer to as the Parallel Postulate, was long suspected by the mathematical community of being provable from the others; indeed, an equivalent of the converse of the Parallel Postulate is proved by Euclid himself in Book I's Proposition 27. Proofs of the Parallel Postulate from the earlier four were attempted repeatedly from ancient times until the nineteenth century. Since the Parallel Postulate really is independent of the other axioms of Euclidean geometry,

[13] One might suppose that general statements are all and only those that begin with a universal quantifier. But as logicians well know, any statement is equivalent to a universally quantified one; for example, p is logically equivalent to $\forall x(x = x \land p)$.

[14] This raises the question of what general vocabulary is. A rough elucidation is that it is vocabulary that applies to all or most entities in the domain or is defined in terms of such vocabulary. Since the task before us is not to define *EP-General* in non-circular terms but to assess it, we shall not dwell on this further.

[15] To be clear, *EP-Independence* is assessed at the level of particular axiomatic presentations. We'll consider presentations that are (or have been) favoured by actual mathematicians. For example, we take it to be a significant discovery that the five standard axioms of Euclidean geometry (see §1) are mutually independent, despite the fact that the theory can be formulated with a single conjunction that obviously satisfies *EP-Independence*.

these 'proofs' were invariably fallacious, or else involved an assumption logically equivalent to the postulate itself.[16]

The history of the issue makes it clear that the question of the independence of axioms was taken to be of central *mathematical* importance. Hilbert, for instance, declares the independence of the Parallel Postulate one of 'the most important results of geometrical enquiries' (1899b: 38). But to qualify as a constituent principle of the EP, the independence of the axioms would additionally have to be of *epistemological* significance for writers in that tradition. Our view is that the question of independence was generally taken to be of epistemological importance, and not just by thinkers in the Euclidean tradition.

The mathematicians attempting throughout the ages to prove the Parallel Postulate were near-unanimous in their agreement that it was true, yet a proof was still sought. For example, writing in the fifth century, Proclus declares that the postulate 'ought to be struck from the postulates altogether' (1970: 150) despite acknowledging that its truth is obvious. This is because 'its obvious character does not appear independently of demonstration but is turned by proof into a matter of knowledge' (1970: 151). This suggests an epistemological principle in the background, for example that that which admits of proof requires it for the highest standard of knowledge – and this would imply a principle such as *EP-Independence*.[17] There is indeed some evidence of such a requirement throughout the history of mathematics. For example, Euclid himself gives proofs for propositions whose self-evidence seemingly outstrips that of the Parallel Postulate; an example is Proposition 20 of Book I, that the sum of any two sides of a triangle is greater than the third. That such propositions are proved, rather than taken as redundant axioms, suggests, although not conclusively, the working of a principle such as *EP-Independence*.

Moreover, it is unclear exactly how the issue of the Parallel Postulate could have come to be of such central importance if its proof had been of merely mathematical interest. Even towards the end of the eighteenth century, the issue was amongst the most important in mathematics, and its unresolved status was declared by d'Alembert as 'the scandal of elementary geometry' (Lewis 1920: 18). This language suggests that the mutual independence of the axioms was not generally thought to have the status of a mere open problem, but rather was of more fundamental importance. So, although interest in independence is not found only in the work of proponents of the EP, *EP-Independence* merits

[16] Lewis (1920: 17–18).

[17] Such a principle is claimed to be 'in the nature of mathematics' by Frege, who also detects the principle at work in Euclid (1884/1953: §2).

inclusion in the programme, at least as one of the four principles supplementary to the three core ones.[18]

EP-Completeness says that the rules and axioms are sufficient for inferring *all* truths in some important class. This, again, is a schematic requirement. Its importance and plausibility depend on the class of truths to be specified. The weakest such principle of any interest is with respect to the *known* truths of the domain. On this version of the principle, the axioms might be viewed as primarily 'an organization of our knowledge, making it more manageable and more interesting' (Russell 1907: 580). The goal would then be to lay out what we already know in geometry, arithmetic, etc., in the most systematic and concise way possible. If *every* known proposition of the relevant science can be inferred from the axioms, the latter have done their job.

A more ambitious version of the principle, which we detect in most of the prominent manifestations of the EP throughout history, is that the axioms must be complete with respect to the *knowable* truths of a particular science. It is not, of course, straightforward to say what knowability amounts to in this context, as the modality packed into it can be understood in different ways. But generally speaking, the Euclidean will claim that any truth which we are capable of knowing in the relevant domain can be known by inferring it from the axioms. If some truth in the domain does not follow from the axioms, either we have more axioms to discover, or it lies forever beyond the bounds of human knowledge. The strongest version of the principle says that *any* truth of the relevant domain can be attained; in modern terminology, the axioms are negation-complete.[19]

To sum up, some version of *EP-Completeness* has traditionally been aspired to, often in one of its more ambitious forms. Despite its prevalence, we include it as a subsidiary principle because different versions of it differ significantly in how demanding they are.

A principle that we do not wish to build into the EP, either in its core or on its periphery, is the metaphysical dependency of theorems on axioms, and more generally of derived theorems on the earlier theorems they are derived from. This idea is very much part of the foundational method as conceived of by Frege, who wrote that the aim of proof was 'to afford us insight into the dependence of truths upon one another' (1884/1953: §2). Frege believed that this dependence (*Abhängigkeit*) was an objective matter, and he was not alone in this. He cites Leibniz as a precursor (1884/1953: §17), and in fact the idea of the

[18] Note that *EP-Independence* is satisfied by an axiomatic presentation if the axioms are in fact independent of one another, even if there is no known technique for proving this. A similar comment can be made about our next principle, *EP-Completeness*.

[19] That is, for every sentence φ about the domain, either φ or not-φ is deducible from the axioms.

dependency of theorems on axioms dates back to Aristotle.[20] Precisely because the relation of dependence is metaphysical,[21] we exclude it from consideration. Its inclusion in the EP would add a whole new metaphysical dimension to a principally epistemological story, and in this Element at least, we wish to stick to epistemology as much as reasonably possible.

So much for a first-pass presentation of the EP's principles. In the first of the four historical sections to follow (§3), we briefly distinguish the *Elements* from the EP. We then consider at greater length whether Aristotle's account in the *Posterior Analytics* qualifies as a version of the EP (§4). In §5, we look at the EP's golden age: the seventeenth-century accounts of Descartes in his *Discourse on Method* and of Pascal in *On the Geometric Mind.* We end the historical material with a post-EP, twentieth-century account of 'Descriptive Axiomatisation' (§6). In the second half of the essay (§§7–9), we assess the EP's seven principles against contemporary mathematics.

Of course, not all historical accounts of mathematics fit the mould of the EP, and notable exceptions are found in the work of British empiricists. Berkeley, for instance, advanced an instrumentalist account of geometry, according to which much of classical geometry is literally false.[22] And Mill famously argued that 'the first principles of geometry are results of induction' (1882: Book 2, Chapter IV). Accordingly, the 'peculiar certainty' attributed to the truths of mathematics is an 'illusion' (1882: Book 2, Chapter V). But despite such exceptions, views which conform to the EP are remarkably common in the history of philosophy. Indeed, even among the British empiricists, we find expressions of sympathy with the EP. For instance, Locke writes admiringly of 'the mathematicians, who from very plain and easy beginnings, by gentle degrees, and a continued chain of reasonings, proceed to the discovery and demonstration of truths, that appear at first sight beyond human capacity' (1689/2004: Book IV, Chapter XII, §7).[23] Given the historical prevalence of the EP, we owe the reader some justification for our choice of historical material in §§3–6. Whether the *Elements* itself is a theory in the Euclidean mould (in our sense) is a question too obvious to ignore. Aristotle's is the most influential ancient,

[20] Aristotle clearly believed that some features of a given substance (its accidents) could be explained by some other features of that substance (its essence); one could therefore argue that, for him, some propositions about X depend on other propositions about X. For a review of the notions of dependence and grounding in Aristotle, see Corkum (2016).

[21] As Shapiro (2009: 183) and others have noted.

[22] See Jesseph (2022) for an account of the development of Berkeley's treatment of geometry and a comparison with his view of arithmetic.

[23] The interpretation of Locke's philosophy of mathematics is notoriously difficult, and we do not claim that it fits the mould of the EP. It is nonetheless striking that this particular passage seems to endorse *EP-Truth*, *EP-Flow*, and something close to, though perhaps falling short of, *EP-Self-Evidence*.

or even pre-modern, account of mathematical and scientific method and arguably influenced Euclid himself. The seventeenth-century accounts represent the EP's apogee. Pascal's particular version is recognised as canonical: Lakatos (1962: 5) calls it the *locus classicus* of the EP, with some justice, as we shall see. Finally, Descriptive Axiomatisation represents a fairly standard contemporary mathematical understanding of how axiomatisation should proceed, and so is worth comparing to the EP. Presenting it also smooths the transition to §§7–8, where we assess the EP from a contemporary perspective, and foreshadows the discussion in §9.

We must, inevitably, issue a caveat: it would take many more pages than we have at our disposal to do full justice to the topic of how, say, the EP relates to Aristotle's thought, or to Descartes', or to Pascal's. We limit ourselves to comparing the EP to some specific texts. Our historical survey omits many other accounts – Frege's for example, which we touch on only in passing.[24] The main reason is practical limitation of space; another is that Frege and his successors are writing at a time when the EP has fallen into decline.

3 Before the EP: Euclid

Since the Euclidean Programme takes its name from Euclid, whoever he might have been,[25] we invert the historical order and start with him before examining Aristotle at greater length in the next section. Our focus will be specifically on the *Elements*.

In §2, we spelt out the EP based on Lakatos's characterisation rather than on Euclid's own *Elements* (c. 300 BC). The reason is that the latter approach would have given us very little to go on. For, as many commentators have noted, Euclid offers us no philosophical gloss on his method. Indeed, his comments on his goals and methods are even more spare than what one might find in today's typical formal textbooks, not known for their epistemological asides.[26] Euclid does occasionally use epistemological terms, for instance when he wraps up an argument by saying that, on the basis of what has been proved, it is now 'manifest' (φανερόν) that such and such.[27] Yet such language is no more than suggestive and hardly constitutes a worked-out expression of the EP. Euclid may be one of the most eloquent mathematicians of all time, but philosophically

[24] A detailed investigation of Frege's relationship to the Euclidean method may be found in Shapiro (2009: 175–96).

[25] We know next to nothing about Euclid's life: see Heath (1925 vol. 1: 1–6) for some scant biographical information, little improved on since Heath's day.

[26] As Davey (2021: 54) observes.

[27] *Passim*, for example, in the corollary to Proposition 16 of Book XIII.

he is reticent. The EP is inspired by the *Elements*, but not by its author's philosophical gloss on what he's up to there.

By the same token, it should not be assumed that the *Elements* conforms to Euclidean precepts. The EP is a methodological ideal inspired by the *Elements*, which until the second half of the nineteenth century was widely believed to instantiate it. But just because an ideal is inspired by something does not mean that it closely corresponds to it. Our Western ideal of democracy was inspired by classical Athenian democracy, whose flaws are now manifest.[28] Similarly, it turns out, there is a gap between the epistemic ideal inspired by the *Elements* – the EP – and the *Elements* itself, as we shall now observe. Duly forewarned, let's consider the EP's three core principles.

3.1 EP-Truth

An important observation is that many of the propositions in the *Elements* (about a third) are imperatival in sense.[29] Some of the postulates (see §1) use a form of the verb 'to request', translatable as 'let it be postulated that', to set up a pretence that there is 'a list of requests not only made but also granted'.[30] This grammatical feature reflects a major difference in how the Greeks understood geometry and how it was later understood, including by us now. The propositions proved by Euclid may be divided into two groups. The first group consists of 'problems' formulated by infinitives, for example the equilateral triangle construction at the very start of Book I (Proposition 1). The second group consists of 'theorems' such as Pythagoras's Theorem, the penultimate one (Proposition 47) of Book I. Although Euclid does not himself use the language of 'problems' vs 'theorems', many other Greek geometers did so to denote the difference between a construction or act on the one hand and a timeless fact on the other. Heath's translation respects this grammatical and semantical difference:

> On a given finite straight line to construct an equilateral triangle. (Proposition 1, Book I)

> In right-angled triangles the square on the side subtending the right angle is equal to the squares on the sides containing the right angle. (Proposition 47, Book I)

This distinction between 'problems' and 'theorems' stands in the way of a blanket attribution of *EP-Truth*. Indeed, it seems to be a category mistake to

---

[28] To state an obvious but profound one: its disenfranchisement of women, slaves (including freed slaves), and metics (foreign residents).

[29] Thanks to Ben Morison for the fractional estimate here and for stressing the distinction between problems and theorems.

[30] Denyer (2022: 4). The postulates are a list of subordinate clauses governed by ᾐτήσθω, the perfect passive imperative of the verb αἰτέω, meaning 'to request'.

ascribe truth to something with the grammatical form of an imperatival axiom. In contrast to Euclid, of course, the EP conceives of *all* axioms and theorems as straightforwardly declarative.

As for Euclid's declarative (non-imperatival) axioms, the majority of them are straightforwardly true. But not all. Today, we would regard the fifth postulate as true of Euclidean spaces, by definition, but not true outright. It fails in non-Euclidean spaces, which are mathematically on a par with Euclidean ones. And as far as physics is concerned, the geometry of actual space is non-Euclidean. So it now seems that the Parallel Postulate is neither self-evident, nor even true, in the straightforward sense that was previously maintained by advocates of the EP. At best, the Parallel Postulate is a stipulation that the geometry we are interested in is Euclidean. We argue below (§§7.1–7.2) that stipulative axioms are indeed self-evidently true, but the failure of the Parallel Postulate in non-Euclidean spaces certainly complicates the ascription of truth (and a fortiori of self-evidence) to this axiom, as it has been traditionally understood.

Likewise, in light of modern set theory, it is natural for us to regard the fifth common notion – that the whole is greater than the part – as false. This is because any infinite set has proper subsets that are the same size as it. A reply might be that this is not what 'greater' should mean here; if by 'greater' is meant 'proper superset' then the common notion is respected, since any set, infinite or finite, is a proper superset of any of its proper subsets (this being the corresponding interpretation of 'part'). But the point is that on the prevailing contemporary reading of 'greater than', Book I's fifth common notion is false. A similar point applies to the third common notion.

3.2 EP-Self-Evidence

Ascribing self-evidence to the permissibility of a construction such as that of an equilateral triangle would be a category mistake. Yet suppose we do consider, anachronistically, non-dynamic versions of Euclid's axioms (postulates and common notions). Let's say we 'propositionalise' their content and turn constructions into timeless propositions of the sort modern mathematics trades in, replacing the text's dynamic idiom with static vocabulary.[31] This, as just explained, would constitute a significant departure from Euclid's (and all ancient Greek geometers') practice. That noted, we can still ask the question: would the resulting propositions be self-evident?

Here we must distinguish two perspectives: Euclid's and ours. Since this is such an anachronistic question – precisely because the propositions have been transmuted from dynamic to static form – it would be hard to deliver a confident

[31] As famously done in Hilbert (1899a).

'yes' from Euclid's perspective. At any rate, such an answer can only be very speculative (though perhaps evidence from other parts of ancient Greek mathematics could help). What about from our perspective? As we observed in §3.1, some of them don't even seem true and so are far from seeming self-evident.

So, to summarise §3.1 and §3.2, there is a categorical problem – because of the issue of dynamic language (and the corresponding presence of 'problems' in Greek geometry as opposed to 'theorems') – in ascribing either *EP-Truth* or *EP-Self-Evidence* to Euclid's starting points. Additionally, from a modern point of view, at least a handful of the statically phrased axioms seem neither straightforwardly true nor self-evident (to us).[32]

3.3 EP-Flow

If one takes the relevant inference relation between propositions of the system to be logical deduction, the *Elements* fails to live up to the standards of the EP. In fact, one hardly needs to go far into it before stumbling upon a logical gap in reasoning. The very first proposition of the *Elements* concerns the construction of an equilateral triangle with a given line segment AB as its side. A typical example of a Euclidean construction problem, it allows the use of a ruler, with which to draw a straight line between two points, as well as the use of a compass, with which to draw a circle with given centre and given radius.[33] In his argument, Euclid draws a circle with radius of length AB centred on A and then a circle with radius of length AB centred on B. At this stage in the reasoning, Euclid (without saying as much) assumes that the two circles meet. This circle-circle intersection property may have been entirely obvious to Euclid and to anyone reading him, but it simply does not follow from any of his postulates or common notions.

Euclid, it turns out, also assumes the method of superposition, which allows him to move one figure and superpose it on another. Various other principles are tacitly assumed in the *Elements*, for example, betweenness facts about points on a line. Some of Euclid's definitions also leave much to be desired from a contemporary perspective. Some of them, such as the tenth definition in Book I, which tells us what a right angle is,[34] are perfectly acceptable. Others, notably the first three of the same book – 'a *point* is that which has no part', 'a *line* is breadthless length', and 'the extremities of a line are points' – do not pin

[32] If other works of Euclid's are included, the moral is even clearer. *EP-Self-Evidence* does not apply, for example, to the basic principles of his *Optics*.

[33] Such constructions are therefore often known as 'compass and straightedge constructions'.

[34] 'When a straight line set up on a straight line makes the adjacent angles equal to one another, each of the equal angles is *right*, and the straight line standing on the other is called a *perpendicular* to that on which it stands.'

down their *definienda* precisely, since they appeal to a pre-existing understanding of the notions to be defined. We would now think of them as mental crutches rather than definitions: too vague to be enshrined in an official account, although they might usefully accompany one. In fact, those opening definitions don't appear to be used in any of the subsequent proofs; whatever they are, they are not there for the purpose of deriving theorems.

By late antiquity – certainly by the time of the fifth-century Proclus – several of the gaps and hidden assumptions in Euclid's reasoning were already known. And gaps, implicit assumptions, and imprecise definitions are, by modern standards, out-and-out deficiencies. Although Euclid had been lauded, indeed quasi-apotheosised by some, for his rigour for more than two millennia, by the end of the nineteenth century it had become apparent that the *Elements*, far from being a paragon of rigour, was in fact woefully unrigorous when judged against the standards that emerged over the course of that transformative century.[35]

But as we explained in §2, we allow inference in a Euclidean system to be topic-specific. It is clear that Euclid appeals to spatial intuition, and the *Elements*, like any other text of Greek geometry, indispensably relies on drawing diagrams for its cogency. Although Euclid does not list all the principles he relies on, once one allows spatial intuition assisted by the drawing of diagrams as a resource, his reasoning is close to watertight. So, our view is definitely not the late nineteenth-century one that the *Elements* is defective as a Euclidean system because its inferences are not all logically watertight. That said, we do think it fails *EP-Flow*, but for the general reason we articulate in §7.3, that *EP-Flow* is an unattainable ideal for beings like us.

So much for the EP's three core principles. Of the peripheral principles, *EP-Finite*, *EP-General*, and *EP-Independence* are satisfied, the potentially problematic cases of the fourth and fifth postulates having been discussed earlier. But the *Elements* fails to satisfy at least the strongest version of *EP-Completeness* (see §8.4). There is no space here to discuss these four subsidiary principles in detail, so instead we reiterate a key point. Whether it instantiates the EP only partly or not at all, the *Elements* stands its ground as a hugely significant piece of mathematics. We should not conflate an ideal (EP) with a putative instantiation of it (the *Elements*), not even one that inspired it.

[35] Pasch (1882) enumerated many examples of gaps in Euclid's reasoning and listed various principles Euclid tacitly relied on. As a result, these supplementary principles are often known as 'Pasch axioms'.

4 Before the EP: Aristotle[36]

The *Posterior Analytics* presents Aristotle's account of axiomatised science.[37] It predates Euclid's *Elements* by half a century or so, which makes it probable that Euclid was taught by at least some contemporaries of Aristotle. How much, if at all, Aristotle influenced Euclid remains unclear.[38] What is undoubtable is that Aristotle greatly influenced later thinkers, including many mediaeval philosophers.[39]

Our question is whether Aristotle's theory of science in the *Posterior Analytics* conforms to the Euclidean Programme. It's hard to give a straight answer, because, as we shall see, there is evidence for a broadly affirmative one as well as for a more negative one.

Let's begin with what Aristotle thinks a mathematical proof should look like. It corresponds to what he calls a demonstration (ἀπόδειξις), as opposed to a deduction (συλλογισμός). He distinguishes between the two and characterises the first – and more important of the two for our purposes – in this passage:

> By a demonstration, I mean a scientific deduction; and by scientific I mean a deduction by possessing which we understand something. If to understand something is what we have posited it to be, then demonstrative understanding in particular must proceed from items which are true and primitive and immediate and more familiar than and prior to and explanatory of the conclusions [...] There can be a deduction even if these conditions are not met, but there cannot be a demonstration – for it will not bring about understanding. (*Posterior Analytics* I.2 71b17–25, tr. Barnes 1993: 2–3)

So, a proof in the EP mould more closely corresponds to a demonstration than to a deduction. A demonstration gives us *episteme* (ἐπιστήμη) of the theorem

[36] We have used the Loeb texts cited in the bibliography for Aristotle's texts in Greek. Translations are as indicated, otherwise by Paseau.

[37] See for example Barnes (1993: xx). In line with Aristotle's thinking, science is here understood in a broad sense to include mathematics.

[38] Ross (1957: 56) argues on internal evidence that the *Posterior Analytics* probably influenced Euclid. In the same passage, Ross also mentions several pre-Euclidean writers of texts known as the *Elements of Geometry* (about whose contents we have no details). A strong assertion of kinship between the *Posterior Analytics* and the geometrical practice of its day is advanced by H. D. P. Lee: 'It is therefore evident that Aristotle's account of the first principles of science in the *Posterior Analytics* is an account of the first principles of geometry; and consequently that it is to geometry he looks to provide the model to which a science should conform. Only on this supposition can we account for the close parallel between the first principles in the *Analytics* and in the Euclidean geometry. It seems quite likely that Aristotle's account of first principles differed little from the account generally accepted by geometers of his day' (1935: 117–8).

[39] Such as al-Fārābī and al-Ghazālī (Pasnau 2017: 28), as well as Christian and Jewish philosophers.

thereby proved, variously translated as 'understanding' or 'scientific knowledge'.[40] We now compare this account with the EP principle by principle, varying the order of the principles.

4.1 EP-Self-Evidence

The word 'axiom' in the EP means an unproved starting point of mathematical inquiry. A complication in comparing Aristotle's methodology to the EP arises in that he distinguishes three different starting points of demonstration and thereby of mathematical inquiry: axioms, definitions, and hypotheses. But as this is of secondary importance, we can think of all three in an undifferentiated way as first principles.[41]

The question before us is whether Aristotle thought that the first principles of mathematical inquiry are self-evident. Several commentators return an unambiguous 'yes'.[42] In its favour is the fact that Aristotle characterises them using the expression δῆλον ἐξ αὐτῶν, word-for-word translatable as 'clear from themselves'; see for example *Topics* I.5 102a11. To probe this further, let's return to the key passage from the *Posterior Analytics* quoted earlier (I.2 71b17–25). In it, one finds axioms in our modern sense – first principles of mathematical inquiry – characterised by Aristotle as claims that are (i) true, (ii) primitive, (iii) immediate, (iv) more familiar than the conclusions, (v) prior to the conclusions, and (vi) explanatory of the conclusions. The six properties are connected in various ways explored in the literature.[43] These sorts of characterisations are not limited to the *Analytics*, *Posterior* or *Prior*. At the start of the *Topics* (I.I), for instance, he defines demonstration as reasoning from premises that are primary and true, such premises being claims that 'command belief through themselves and not through anything else'.[44]

[40] For 'understanding' see Burnyeat (1981) and for 'scientific knowledge' Bronstein (2016: 18–20).

[41] The first type is what he calls axioms (ἀξιώματα) or first principles (κοινά or κοιναὶ ἀρχαί). These are highly general principles presupposed by the very possibility of knowing anything, or at least principles whose scope goes beyond that of a single science. Euclid in the *Elements* calls them common notions. Aristotle mentions the law of non-contradiction and the law of excluded middle as examples, which we think of as logical principles; but he also includes principles which for us are of lesser generality, for example, that if equals be taken from equals then equals remain (see Ross 1957: 55 for several citations). The second are what he calls definitions (ὁρισμοί), which state the meaning of a term, and the third element of the triad hypotheses (ὑποθέσεις): existential claims more specific to the subject at hand. There is a close correspondence between Aristotle's hypotheses and Euclid's postulates, which, as we have seen, affirm the possibility of constructions. As well as *Posterior Analytics* I.2 72a7–25, we have also drawn on Lee (1935) here.

[42] Examples include Heath (1925, vol. 1: 124) and Lloyd (2014: 16).

[43] For example, in Barnes (1993: 93–6). Following Barnes, Detel (2012: 258) takes immediacy and explanatory power to imply the other four properties.

[44] ἔστι δὲ ἀληθῆ μὲν καὶ πρῶτα τὰ μὴ δι' ἑτέρων ἀλλὰ δι' αὑτῶν ἔχοντα τὴν πίστην (I.1 100b18).

All this suggests that Aristotle subscribes to *EP-Self-Evidence*. But there is a rival interpretation, which sees Aristotle as engaged in a different sort of project. A few lines below the passage containing the sixfold characterisation of first principles (I.2 71b33), he explicates the condition of being 'prior and better known' by distinguishing two senses of this last phrase. One is that the principles be 'better known to us' (a sense very congenial to reading Aristotle as a proto EP-theorist); the other is that they be 'more knowable by nature', which might be glossed as 'more knowable in the nature of things'. In the second and apparently operative sense, first principles are first in the order of explanation and the source of knowledge of everything else. Although such propositions may be called self-explanatory, crucially, they need not and typically will not be self-evident (Barnes 1993: 97); we have to discover them and work hard to understand them. For Aristotle, they will mostly be definitions, and a glance at some of these definitions suggests that they are far from self-evident. An example from the *Physics*: 'change is the actuality of that which exists potentially, in so far as it is potentially this actuality' (III.1 201a, tr. Waterfield 2008: 57).[45]

We cannot adjudicate between these two pictures, currently disputed by Aristotle scholars.[46] But we note that the issue crucially affects whether we see Aristotle as a forerunner of Euclidean foundationalism or not. On the first picture, first principles are self-evident or close enough; on the second, they are starting points of explanation, with no presumption of self-evidence.

4.2 EP-Flow

Which of the two pictures articulated in §4.1 is the more appropriate one also affects whether we attribute some version of *EP-Flow* to Aristotle.

On the first picture, construing *episteme* as scientific knowledge, we gain this sort of knowledge by discovering first principles and then inferring conclusions from them via demonstrative syllogisms (*Posterior Analytics* I.2). We stand in a privileged epistemic relation to first principles, and working through a demonstration from them to theorems allows us to attain scientific knowledge of the latter. Geometry and arithmetic are instances of this method at work (*Posterior Analytics* I.7–10). So, on this picture, the weak version of *EP-Flow* may be attributed to Aristotle: any subject who bears relation E to premises to a high degree and clearly grasps that the premises entail the conclusion then bears relation E to the conclusion to a fairly high degree.

[45] We are particularly indebted to Ben Morison in this paragraph. We set aside the question of whether, for Aristotle, definitions are real definitions, giving the essence of something.

[46] See Zuppolini (2020), who calls the first the rationalist account and the second the interrelational one.

On the second and rival picture mentioned in §4.1, in contrast, when you see that q because p, you grasp that p is prior to q in the order of explanation. This does not come with the transfer of some of the more obvious epistemic goods. In particular, explanatoriness cannot be the epistemic good transmitted from axioms to theorems, since at least in principle a theorem might not be explanatory of anything. The gerrymandered good 'grasping the proposition's place in the explanatory network' is a candidate, but it is not quite the sort of good the EP-theorist has in mind. Perhaps a better candidate is understanding: when you trace q's explanation back to first principles you understand q better.[47] This chimes with the case made by Aristotle's more recent interpreters that *episteme* is best translated as 'understanding', where understanding the principles of science raises one to the highest epistemic state.

But even if we take understanding to be the relevant good, the ascription of *EP-Flow* to Aristotle on this second picture is complicated, because it's not clear that there is any sort of temporal priority here. Perhaps light dawns on the whole when you see the explanatory connections from axioms to theorems.[48] If so, Aristotle's account is holistic rather than sequential. Interestingly, this would bring it closer to the extrinsic-justification story we will briefly mention later (in §9), whereby some axioms earn their keep by dint of what they allow us to prove.

In short, the same sort of reading of Aristotle that sees him as subscribing to *EP-Self-Evidence* also has him subscribing to at least a weak version of *EP-Flow*. The alternative reading is much less clearly compatible with a version of *EP-Flow*.

4.3 EP-Truth

As the passages quoted earlier and many others show, Aristotle unambiguously ascribes truth to first principles. This is the least controversial part of the correspondence between Aristotle's philosophy of science and the EP.[49]

The theorems of a given science must also be true on Aristotle's account. Although he does not present a logical system with a clear distinction between axioms on the one hand and rules of inference on the other, what we would now recognise as the rules of inference he employs are truth-preserving. The most compelling modern reconstructions of Aristotle's syllogistic logic are those of Corcoran (1974) and Smiley (1973). Both take the syllogistic to be a

[47] For Aristotle and understanding, see Burnyeat (1981); section 1 of Morison (2019) offers a summary.

[48] As suggested in section 6 of Morison (2019).

[49] The Aristotelian corpus also contains some remarks on the nature of truth; see Crivelli (2004) for a recent book-length investigation. We distinguish Aristotle's account of truth from his common-or-garden uses of the word.

natural-deduction system with inferentially primitive inference rules. In Smiley's version (1973: 141), these rules are:

from 'All *A* are *B*' and 'All *B* are *C*', infer 'All *A* are *C*';
from 'All *A* are *B*' and 'No *B* is *C*', infer 'No *A* is *C*';
from 'No *B* is *A*', infer 'No *A* is *B*';
from 'All *B* are *A*', infer 'Some *A* is *B*'.

These rules are all obviously truth-preserving and thus sound, as Aristotle recognised. (For the last rule, note that Aristotelian predicates are assumed to have non-empty extensions.)

4.4 EP-Finite

Assuming that a natural-deduction account à la Corcoran/Smiley is the best reconstruction of Aristotle's syllogistic, Aristotle can be interpreted as believing that the rules of inference are finitely many. As for the axioms of total science, Aristotle unambiguously states in *Posterior Analytics* I.32 88b6 that science reaches an infinite amount of conclusions, and he observes in the next but one sentence that it could not do so if the set of principles were finite. If we take Aristotle's explicit assertion at face value, we may read him as denying that the set of axioms is finite. In the absence of any particular reason to exempt mathematics from this verdict, it applies to it too.[50]

4.5 EP-General

In his theory of science, Aristotle focuses on explananda – conclusions of demonstrations – that are universal propositions, for example, 'All statues made of metal are heavy'. In these cases, the premises of the demonstration will also be universal; for the given example, one of the premises (the so-called major premise) would be 'All things made of bronze are heavy' and the other (the minor premise) 'All statues made of metal are made of bronze', together making up an instance of the Barbara syllogism. The premises of such demonstrations are general. This, for Aristotle, is a paradigm case of demonstration.

Nonetheless, Aristotle allows that there may be demonstrations of particular facts as well as universal ones.[51] One of his examples is 'The Persians waged war on the Athenians'.[52] The premises of a demonstration of this fact cannot be entirely made up of universal propositions, since syllogistic

[50] For more on Aristotle's discussion of these points in *Posterior Analytics* I.32, see Barnes (1993: 194–8).

[51] *Prior Analytics* I.33 47b21–34 and II.27 70a16–20; *Posterior Analytics* I.24 85a14.

[52] A free translation of *Posterior Analytics* II.11 94a37–38.

inferences from universal facts cannot result in a particular proposition such as this one. As he acknowledges, it must include particular premises, such as the fact that the Athenians raided Sardis, which began hostilities and provoked the Persians' retaliation.[53]

The same applies more specifically to mathematics and its axioms. Aristotle's syllogistic is supposed to be the logic of mathematical demonstration. As he puts it when considering the first syllogistic figure:

> Of the figures, the first is especially scientific. The mathematical sciences carry out their demonstrations through it – e.g. arithmetic and geometry and optics. (*Posterior Analytics* I.13 79a18–20, tr. Barnes 1993: 22)

But in syllogistic logic, singular facts cannot be demonstrated from purely universal premises. The contrast is with modern systems of logic, which contain rules of inference taking universal premises to particular conclusions, the prime example being the universal-instantiation rule (from $\forall xFx$ infer Fa). It follows that certain particular facts must be undemonstrated first principles if conclusions of particular facts are to be reached. The assumption being, of course, that valid arguments from universal premises result in universal conclusions.

All that said, Aristotle thinks of demonstrations of universal facts as more scientific and superior to demonstrations of particular facts.[54] As far as this superior type of demonstration goes, then, Aristotle unequivocally subscribes to *EP-General*. But that is in a sense baked into his theory of science, since better demonstrations are of universal propositions and hence must be traced back to universal premises.

4.6 EP-Independence

This is not something which Aristotle comments on directly, to the best of our knowledge. Indeed, Detel (2012: 257) observes that Aristotle did not uphold the later ideal of 'compress[ing] the content of a whole theory into as few as axioms as possible'. Rather, an Aristotelian axiomatisation strives to analyse a scientific theory's content and thereby enables us to understand it better.

Nevertheless, *EP-Independence* is entirely consonant with some sort of asymmetry between axioms and theorems in Aristotle's account. The requirements that the axioms be *primitive*, *immediate*, *prior* to the theorems, and that they 'command belief through themselves *and not through anything else*' (our italics) would heavily suggest that an axiom derivable from the remainder of the theory be disqualified as such. We therefore suggest that *EP-Independence* is

[53] *Posterior Analytics* II.11 94b1–8. [54] *Posterior Analytics* I.14.

congenial to Aristotle's account, without ascribing to him a full commitment (explicit or tacit) to such a principle.[55]

4.7 EP-Completeness

If Aristotle's system is recast in natural-deduction terms, its rules are complete with respect to syllogistic inference (Smiley 1973: 141). But are they complete for the purposes of science? The textual evidence is unclear. Aristotle plainly took his syllogistic to be the underlying logic of science, which is why the *Posterior Analytics* builds on the *Prior Analytics*, as the names indicate, and as Aristotle tells us himself.[56] So, one could maintain on this basis that Aristotle accepted *EP-Completeness* in some form and believed that his syllogistic fits the bill. But in fact, Aristotle's syllogistic falls far short of the logic required even for the known mathematics of his time.[57] From our vantage point, this is easy to see, and we need look no further than the first proposition of Euclid's *Elements* about the construction of an equilateral triangle from a segment, discussed in §3 earlier. Any formalisation of this proof requires the resources of at least first-order logic.[58] Naturally, this leaves us with an exegetical question: did Aristotle really not appreciate that the logic of mathematics went beyond his syllogistic? Since it is tangential to our purposes, we raise the question only to shelve it. The point for our purposes is that mathematical theories in the form envisioned by Aristotle, that is, those whose underlying logic is syllogistic, will fail to satisfy even the weakest form of *EP-Completeness*.

That concludes our discussion of the three core EP-principles and the four subsidiary ones. We have, of course, no more than scratched the surface of Aristotle's ideas relevant to the EP, even restricting attention to the *Posterior Analytics*. The task of relating Aristotle more precisely to the EP is a difficult and ambush-filled one, where specialists fear to tread and non-specialists are well-advised not to rush in. There are clearly some commonalities between Aristotle's philosophy of science and the EP, although we also mentioned some

[55] A complication arises from the way that theorems of one science may be used as axioms in another, as when someone doing optics can just use without proving again theorems proved in geometry. That is the main way we can envisage Aristotle allowing a theorem to be an axiom.

[56] See for example *Prior Analytics* I.4 25b26–31.

[57] Under the (safe) assumption that the mathematics Aristotle was acquainted with was not too dissimilar from the mathematics Euclid systematised not long after Aristotle's death.

[58] Note, for example, the proof's construction of specific points. In this regard, Ian Mueller remarks: 'I have analysed *Elements* I,1 in order to show that Euclid's tacit logic is at least the first order predicate calculus, nothing less. His logic may be even more than that, since representing his reasoning in the first order predicate calculus would seem to require reformulations foreign to the spirit of the *Elements*' (1974: 43).

reasons not to construe Aristotle as a proto EP-proponent. We now turn to more full-fledged early modern expressions of the EP.

5 The EP's Seventeenth-Century Apogee

Just under two millennia separate Euclid from the great thinkers of the seventeenth century. During that time, numerous treatises modelled on, influenced by or simply about the *Elements* were written, many no longer extant. A great many mathematicians and philosophers took their point of departure from the *Elements*, not just in the West but in the Islamic world as well. We know, for instance, of more than fifty commentaries on the *Elements* written by Islamic mediaeval thinkers.[59] And we also know of the rich mathematical work on Euclid by Islamic mathematicians that pre-empted ideas which surfaced many centuries later in Europe.[60]

The seventeenth century in particular saw an explosion of interest in the Euclidean method.[61] Wardhaugh comments that seventeenth-century European thinkers 'use[d] Euclid's *Elements* as a specimen of how thought should be done, how knowledge was structured or how reason really worked' (2020: 151). He notes the titles of the following works of that period: *The Euclid of Logic*, *The Euclid of Medicine*, *Elements of Jurisprudence*, and *Elements of Theology*. He also points out that there were earlier precedents for many of these, including Proclus's fifth-century *Elements of Theology*. The best-known instances of this seventeenth-century trend are perhaps Newton's *Principia* and Spinoza's *Ethics*, or to give the latter its full Latin title: *Ethica Ordine Geometrica Demonstrata* ('Ethics Proved in The Geometrical Manner'). As the title indicates, Spinoza's treatise attempted to impose a Euclidean deductive structure on a philosophical subject matter. For Spinoza, the Euclidean method is the paradigm of certain knowledge and immune to doubt.[62] Newton's *Principia* proceeds in a similar and very deliberate Euclidean spirit. It starts off from some axioms – Newton's laws – and then strives to derive all other propositions from these via logic and mathematical analysis. Of course, as Newton is well aware, his laws are justified not in the Euclidean manner, but empirically.

[59] Freudenthal (1988: 106–7).

[60] For example, in Alhazen's *Solution of the Difficulties of Euclid's Elements* (*Ḥall shukūk fī Kitāb Uqlīdis*), in which he attempts to prove the Parallel Postulate by assuming its negation and proceeding by *reductio*.

[61] Traditional Aristotelian logic continued to be taught well into the heyday of the EP, but from the fifteenth century onwards, the rediscovery of Euclid in Europe facilitated a shift towards regarding the *Elements* as the paradigm of rigorous thought in general (Mancosu and Mugnai 2023: 34–5).

[62] See Craig (1996: 44–51) for more on the relationship between Spinoza's ethics and the Euclidean ideal. Craig also observes (1996: 20) that mathematical method was given pride of place in seventeenth-century philosophical thought more generally.

This section will focus on two French seventeenth-century advocates of the Euclidean method: Descartes and Pascal.

5.1 Descartes

Descartes acknowledges the Euclidean method as the driver behind his own philosophy. We cannot hope to do justice here to Descartes' entire corpus concerning methodology, or even geometry, but instead focus on the most important source where the two subjects intersect. In the *Discourse on Method*, Descartes gives an epistemological programme for seeking 'truth in the sciences' (as the subtitle puts it), casting his net far wider than geometry, although certainly including it. The epistemological notion that Descartes aims to characterise in this text (and in his writings more broadly) is best thought of not as knowledge in the everyday sense, but rather *scientia*, the Latin successor to Aristotle's notion of *episteme* (§4). This is not just knowledge *that* certain truths are the case, but also knowledge of *why* they are the case; it is systematic understanding of a subject matter built on foundations so solid that they do not subsequently admit of rational doubt.[63]

When it comes to *scientia*, Descartes is, of course, a foundationalist. Knowledge is to be first obtained through truths apprehended clearly and distinctly, which cannot rationally be doubted, and from which subsequent knowledge is to be deduced by an infallible course of reasoning. In mathematics, Descartes' approach is simply an instance of this more general foundationalist attitude. Indeed, foundationalism in mathematics is the inspiration for Descartes' more general account, a debt acknowledged early in the *Discourse*[64]:

> These long chains of reasonings, so simple and so easy, which the geometers customarily used in order to arrive at their most difficult demonstrations, had given me occasion to imagine that all things that can be understood by men follow from one another in the same way [...] (1637/2001: 16–17; AT 6: 19)

Scientia is not easy to obtain; for Descartes, as for Aristotle, the account is of a normative epistemic ideal, the pinnacle of possible achievement for creatures of our finite mental powers.[65] Yet although *scientia* is difficult to achieve, it is not impossible, according to Descartes. He even claims to have achieved it himself in a number of domains; the *Discourse* was published with appendices

[63] Sorell (2016: 423–4).

[64] For works of Descartes, two citations are provided. The first is to the relevant translation. The second is to Adam and Tannery's *Oeuvres de Descartes* in volume: page format, for readers who wish to consult the original texts.

[65] Indeed, as Pasnau notes (2017: 24), it is dubious to regard Descartes as providing an account of 'knowledge' in the modern sense at all. What Descartes gives a theory of is the ideal summit of our possible epistemic achievements.

that sought to demonstrate the method at work in optics, meteorology, and geometry.[66]

The geometric method characterised by Descartes fits our characterisation of the Euclidean Programme rather neatly. His commitment to *EP-Flow* is perhaps the easiest to discern. He instructs us to 'follow the correct order' in our reasoning (1637/2001: 18; AT 6: 21), beginning with the propositions which are 'simplest and easiest to know' (1637/2001: 17; AT 6: 19). The conclusion of a demonstration carried out via this method is 'certain and evident' (1637/2001: 17; AT 6: 19), suggesting that a strong version of *EP-Flow* is adhered to.

In his *Regulae*, Descartes addresses a possible difficulty with particularly long and difficult deductions, wherein 'our intellectual capacity is often insufficient to enable us to encompass them [the propositions involved in the deduction] all in a single intuition' (1984a: 26; AT 10: 389), which might be thought to undermine the commitment to *EP-Flow*. However, he seems to think that these issues relating to memory and intellectual capacity are practical; in principle at least, 'a continuous movement' of thought can be rehearsed to the point where 'memory is left with practically no role to play' and we 'seem to intuit the whole thing at once' (1984a: 25; AT 10: 388). And despite Descartes' attention to these practical issues, he defines a deduction as an 'inference of something as following necessarily from some other propositions which are known with certainty', the upshot being that 'very many facts which are not self-evident are known with certainty' (1984a: 15; AT 10: 369).

Descartes' commitment to *EP-Flow* is perfectly compatible with his hostility to the logic of the time. He writes of the then-current syllogistic logic that 'although it contains, in effect, many very true and good precepts, there are nevertheless so many others mixed among them which are either harmful or superfluous, that it is nearly as hard to separate them as it is to sculpt a Diana or a Minerva out of a block of marble which is still roughhewn' (1637/2011: 15; AT 6: 17). But the principle of *EP-Flow*, as we characterise it, allows for inferences that are not purely logical and need not conform to the dictates of some particular logical system such as Aristotelian syllogistic (as we stressed in §2).

Regarding the other core principles of the EP, we note that the very first injunction of the method in the *Discourse* is to never accept as true a proposition

[66] We might wonder how exactly the geometrical method is supposed to work in optics or meteorology, given that the first principles in these disciplines would presumably fall short of self-evidence. The answer is unclear, and the question seems to have troubled Descartes as well. The *Optics* and *Meteorology* both proceed from what Descartes calls 'hypotheses', rather than self-evident axioms. Descartes describes his hypotheses as 'easy' and 'simple' but admits that he lacks the requisite demonstrations of them (1637/2011: 264; AT 6: 233). Nonetheless, he is insistent that his geometry-inspired method is as useful outside of mathematics as within it (1637/2011: 18; AT 6: 21).

of which one does not have evident knowledge (1637/2011: 16; AT 6: 18). This clearly exhibits Descartes' commitment to a form of *EP-Self-Evidence*. Given Descartes' general foundationalist stance, if all known propositions are evident, then some of them must be self-evident. Or as Descartes puts it, they must present themselves to the mind so clearly and so distinctly as to be beyond doubt. Since what is apprehended clearly and distinctly must be true (a general principle of Descartes' epistemology), and since the result of deduction is also certain and evident for Descartes, we also see here a commitment to *EP-Truth*.

Turning to the EP's subsidiary principles, Descartes' geometric method certainly contains some version of *EP-Completeness*, though it is a little unclear which. At times, Descartes seems to think that the geometric method is complete in the strongest sense, such as when he imagines that by application of the method 'there can be [nothing] so remote that we cannot eventually come upon it, or so hidden that we cannot discover it' (1637/2011: 17; AT 6: 19). However, closer inspection suggests that Descartes' version of *EP-Completeness* is that all *knowable* truths are discoverable by deduction from the axioms. The expression 'all things that can be understood by men follow from one another in the same way' is most naturally understood this way, and Descartes does explicitly give examples of things he considers to be unknowable; for example, 'the ratios between straight and curved lines are unknown, and even, I believe, unknowable to men' (1637/2011: 206; AT 6: 412). Importantly, curves and construction techniques that prevent one from reaching 'exact and assured conclusions', such as lines which are sometimes straight and sometimes curved, are excluded altogether from geometry (1637/2011: 206; AT 6: 412). Thus Descartes imposes a mathematical requirement which guarantees the satisfaction of the philosophical goal of *EP-Completeness*; that which is unknowable by geometric methods is simply excluded from the class of problems with respect to which completeness is sought.[67] In short, we may ascribe to Descartes a version of *EP-Completeness* according to which all that is knowable in geometry can be deduced from the axioms.

The other subsidiary principles are more difficult to assess. There is no obvious demand for the generality of premises in his method, despite his preference for beginning with propositions that are 'the most simple and the most general' (1637/2011: 18; AT 6: 20). This is particularly evident when we look at the application of Descartes' method in areas outside geometry, where we are allegedly able to have clear and distinct knowledge of particular objects. And while, for example, our knowledge that God exists is likened to knowledge of a geometric *theorem* (1637/2011: 30–31; AT 6: 36), other examples, such as in the *cogito* (1637/2011: 27–28; AT 6: 32), seem rather to have the status of

[67] Thanks to Sébastien Maronne for this point.

axioms or first principles. But a proposition concerning the existence of a particular being (Descartes himself) does not have the general character typical of axioms in the EP. We tentatively suggest therefore that Descartes may not be a subscriber to *EP-General*.

The situation with *EP-Independence* and *EP-Finite* is less determinate. There is no clear evidence discernible either for or against the inclusion of either principle in Descartes' version of the EP, though his disparagement of syllogistic logic for including 'superfluous' principles, which we saw earlier, may at least suggest some sympathy with *EP-Independence*.

We can conclude, therefore, that Descartes at least preaches the core elements of the Euclidean Programme in the *Discourse*, even if he does not display a clear commitment to all of its subsidiary aspects. It is, however, immediately noticeable that the *Geometry* does not follow the traditional Euclidean presentation, with definitions, common notions, and postulates given at the start and theorems deduced subsequently. One should not, however, take this to mean that Descartes is not a Euclidean in our sense. The *Geometry* was intended by Descartes to be read by an educated audience,[68] and he explicitly assumes that the reader is familiar with the geometry of his time (AT 6: 368). Moreover, the presentation is explicitly designed to 'cultivate' the mind of his readers by teaching them to solve geometric problems of the kind discussed in the text (1637/2001: 180; AT 6: 374). This gives us a means to explain his decisively un-Euclidean presentation in the *Geometry*. In the later *Meditations* (second set of replies), Descartes remarks that he does not think that a *synthetic* (i.e. Euclidean) presentation of a theory is the best method of teaching it; in this, he is surely correct. Rather, an *analytic* presentation (corresponding roughly to the order of discovery) is the 'best and truest method of instruction' (1984b: 110–11 for the quotation and a helpful explanation of Descartes' terminology; AT 7: 155–6). And in the same work he does give a presentation of some of the central arguments of the *Meditations* in traditional Euclidean style and refers to this as a geometrical presentation. So we should not take the un-Euclidean presentation of the *Geometry* at face value; it is written in that way for a specific purpose relating to a specific audience.

That said, it is questionable whether Descartes' *practice* lives up to his lofty Euclidean aspirations. The third appendix published alongside the *Discourse* is the *Geometry*, which supposedly exhibits the methodology of the *Discourse* in the geometric context. But the actual mathematical work of Descartes' *Geometry* instantiates the ideal of the EP at best only approximately. Despite its clear status as a work of enormous mathematical significance, measured by the standards of the *Discourse*, the *Geometry* fares rather poorly.

[68] Thanks to Sébastien Maronne for highlighting the importance of this point.

The main violation of the EP in Descartes' *Geometry* relates to *EP-Flow*. On numerous occasions, Descartes employs a number of true propositions which have not been proved at the time of their employment. While this does not contradict the generic formulation of *EP-Flow* provided in §2, it does contradict Descartes' own version of it given earlier. Descartes explicitly states in the *Discourse* that he will not accept anything which he has not presented clearly and distinctly (1637/2001: 16; AT 6: 18), which in this context should be only an axiom or something that has been proved already.

For example, consider Descartes' geometric interpretation of multiplication, illustrated by the diagram below. Let AB be of unit length. If we want to multiply BD by BC (which meet at an acute angle), we draw a line between A and C. Then extend the line BC and connect D to the line by drawing a segment parallel to AC. Then by the likeness of triangles, BA:BD = BC:BE. Therefore, the ratio between unity and BD is identical to the ratio between BC and BE, thus BE is the product of BD and BC. (Descartes 1637/2001: 177–8; AT 6: 370).

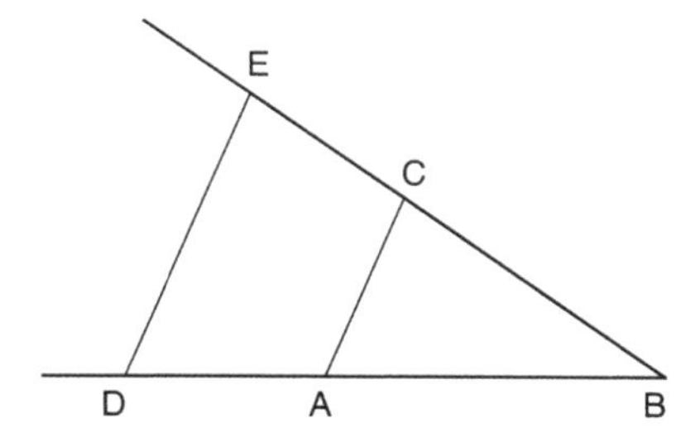

This explicit interpretation does not proceed by methods admissible by the *Discourse*'s standards, as Grosholz highlights (1991: 22–3). In particular, this interpretation of multiplication relies on a theorem about the similarity of triangles. And the interpretation of the arithmetical operations is not an isolated incident. For example, in the very next paragraph, Descartes gives a method for extracting square roots that presupposes at least Pythagoras's theorem.[69] In the next book, Descartes mentions, without proof, 'a general rule for reducing to a cube all the difficulties of the fourth degree' (1637/2001: 195; AT 6: 395–6). According to Bos (1981: 305), this is an allusion to Ferrari's rule for reducing equations of the fourth degree in one unknown to those of third degree. And in Book III, when discussing his own procedures for the reduction of equations, Descartes concludes, without argument, that if a plane problem leads to an equation of third or fourth degree, then the equation is reducible to a quadratic (Bos 2001: 390 and n. 22).

[69] Thanks to Doug Jesseph for this example.

It is not immediately clear, however, that Descartes has sinned against his own version of the Euclidean ideal. As discussed earlier, he deliberately avoids the traditional Euclidean style in the *Geometry*. He could, therefore, be attempting to provide an abbreviated example of the Euclidean ideal at work, where familiar proofs or obvious details are omitted. Indeed, Descartes pleads this himself, writing that 'I have not demonstrated here most of what I have said, because the demonstrations seem to me so simple that, provided you take the pains to see methodically whether I have been mistaken, they will present themselves to you' (1637/2001: 244; AT 6: 464).

But the situation is not so simple. While many of the results relied upon tacitly by Descartes were known to have proofs at the time, the *Geometry* makes free use of unproven mathematical conjectures. The most striking example of this is Descartes' general claim that curves which admit of a point-wise construction can be traced by a regular and continuous motion (1637/2011: 206; AT 6: 412). This was not proved until the nineteenth century, by Kempe (Domski 2009: 123–4 and n. 19). This is not an isolated incident, and nor are the other examples harmless. For instance, Descartes claims that all geometrical curves are the solution to a Pappus problem (1637/2001: 196; AT 6: 397); far from merely lacking a proof, this proposition is in fact false![70] Similarly, consider Descartes' extension of Ferrari's rule to the claim that any equation of the sixth degree can be reduced to an equation of the fifth (1637/2001: 195; AT 6: 395–6). Descartes makes no attempt to prove this, in a clear violation of the maxims of the *Discourse*, and the claim is indeed false (Grosholz 1991: 46).

So, although Descartes gives an identifiable endorsement to the EP, we should not take him to have lived up to his own lofty standards. It is worth pausing to reflect on the significance of this. The mathematics of the *Geometry* does not fulfil the standards Descartes sets for the achievement of *scientia*, but this is, for the most part, a distant epistemic ideal. He himself acknowledges that one can apprehend the truth, even in mathematics, without *scientia*. For example, Descartes concedes that someone without *scientia* could be clearly aware that the angles of a triangle sum to two right angles (1984b: 101; AT 7: 141). Could we defend Descartes on the ground that the *Geometry* is supposed to simply give us mathematical knowledge in an ordinary sense, rather than *scientia*?

It seems to us that the answer must be 'no'. Quite aside from the textual issue that the *Geometry* is supposed to exhibit the method of the *Discourse*, and that said method is aimed at achieving the highest epistemic standards, Descartes subscribes to an additional principle that we might call *EP-Chauvinism*. According to this principle (not one of our core or subsidiary ones), the EP

[70] Thanks to Sébastien Maronne for this example.

describes the *best* possible way of attaining knowledge, at least for creatures like us.[71] Knowledge obtained in any other way is, at best, sub-standard. *EP-Chauvinism* raises some thorny questions, which we leave for another occasion, not least because it is a meta-principle, not on a par with our seven EP-characterising principles: it does not intrinsically describe the EP but implicitly compares it to other ways of attaining knowledge or belief. What is clear is that Descartes is a fervent adherent of *EP-Chauvinism*. He affirms that of 'all those who had hithertofore searched for truth in the sciences, only the mathematicians had been able to discover some demonstrations – that is, some certain and evident reasons' (1637/2001: 17; AT 6: 19). He even goes so far as to claim that 'when we are dealing solely with the contemplation of the truth, surely no one has ever denied that we should refrain from giving assent to matters which we do not perceive with sufficient distinctness' (1984b: 106; AT 7: 149). Put more directly, in the pursuit of knowledge independent of practical considerations, which surely includes pure geometry, only *scientia* is good enough. Hence Descartes gives us a fascinating example exhibiting the vast distance between the demands of Euclideanism and the practice of mathematics, a theme to which we shall return in §§7–8.

5.2 Pascal

Another seventeenth-century writer who offers a particularly crisp formulation of the EP is Blaise Pascal. His *De l'esprit géométrique* ('On the Geometrical Mind', hereafter *Esprit* for short) was written in the 1650s.[72] Never published in Pascal's lifetime, it survived in a copy of the manuscript made by one Louis Périer. A short work, extracts from it were first published several decades after Pascal's death, and to this day no complete English translation seems to have been published.[73]

The *Esprit* may have been intended as introductory material for a broader treatise on geometry by Pascal, which, following criticism by Antoine Arnauld, he committed to the flames. Enormously influential on the Port-Royal Logic,

[71] The chauvinistic attitude appears much earlier than Descartes, however. In Aristotle's epistemology, *episteme* as characterised in the *Posterior Analytics* is thought to be the highest possible state of knowledge, unlike our everyday knowledge which is deficient in various respects (Pasnau 2017: 5).

[72] Mesnard (the editor of Pascal's works) devotes several pages to the question of the work's date of composition (Pascal 1655/1991: 368–76) and settles on 1655.

[73] Translations from the *Esprit* here are by Paseau. Abbreviated page references are to Mesnard's edition (Pascal 1655/1991). There is unfortunately no room to contrast the *Esprit* with Pascal's mathematical writings, as we did with Descartes. As Doug Jesseph pointed out to us, an interesting point to examine would be Pascal's quite casual reliance upon infinitesimal methods, particularly in his examination of the cycloid, given that traditional Euclidean geometry is developed in a finitary way.

which incorporates some material from it almost verbatim and in other places paraphrases it, the *Esprit* introduces the contemporary conception of definition in its reduced, essentially abbreviatory, role. As mentioned in §2, Lakatos calls the *Esprit* the EP's *locus classicus*, a brief and intriguing tip of the hat to Pascal and an inspiration for the present essay. Yet despite this accolade by Lakatos, and the *Esprit*'s influence on the Port-Royal Logic,[74] the philosophical literature in the analytic tradition on it is virtually non-existent.[75]

At the very start of the *Esprit*, in the short introduction prior to Section I (390), Pascal observes that geometry offers the perfect example of how to establish truths in such a way that their proof cannot be overturned (so that the proof is, in the French original, 'invincible'). At the start of Section I (391), he comments that geometry is one of the few human sciences to produce infallible ('infaillibles') proofs. In Section II, he comments that a proof that adheres to the canons he has laid out, such as one might find in geometry, cannot raise the slightest doubt (418). Back in Section I, he claimed that what goes beyond geometry is beyond our ken (393: 'ce qui passe la géométrie nous surpasse'), and that geometry is the only science that exemplifies the correct method of reasoning (391–2). A little later (394–5), he remarks that the ideal method would be to define all one's terms and to prove absolutely everything; but since we cannot do that – we cannot proceed back indefinitely – we must make do with the next best thing. We must resort to primitive undefinable terms and principles so obvious that none more obvious might be used to prove them. This method assumes only principles judged clear and constant by the natural light of reason ('la lumière naturelle'). Trying to further elucidate geometrical primitives would engender more confusion than enlightenment (396).

It is clear from these passages that Pascal subscribes either to *EP-Self-Evidence* or something very much like it, since, for him, axioms are indubitable and knowledge of them is not mediated via any other principles. He even uses the word 'evident' ('évidents') in connection with axioms (418), and in a passage setting out his methodological precepts he urges that axioms which are 'perfectly evident' (420) are not to be questioned, manifestly intending those of geometry to be included under this heading. Pascal, it is true, also cautions that people are prone to mistaking incomprehensibility for falsehood; we wishfully think we have direct access to the truth (404). But if faced with an apparently inconceivable proposition *p*, Pascal urges that, before dismissing it

[74] Acknowledged by the Port-Royal Logic's authors: see Arnauld and Nicole (1683/2019: 21).

[75] More historical approaches are better covered. To name but three secondary texts: Gardies (1982) focuses on definitions; Gardies (1984) considers Pascal within a broader historical context; and chapter 4 of Moriarty (2020) is concerned more generally with Pascal's thought in the *Esprit* and related work.

summarily, we consider its negation carefully. Should not-*p* turn out to be manifestly false, we must then wholeheartedly accept the previously inconceivable *p* (404). Pascal, in other words, recognises that the appearance of veridicality or falsehood can mislead; but with the right method in hand, one can separate the self-evident real McCoy from the fool's gold, as we might put it.

It is equally clear that Pascal takes axioms of the just-mentioned sort to be truths ('vérités'), since he calls them that. So, *EP-Truth* may safely be attributed to him. Furthermore, he thinks that proving a theorem from the axioms establishes its epistemic credentials, impeccably so. This follows from one of the *Esprit*'s earliest lines, in which Pascal speaks of proving *already discovered* truths (390: 'véritées déjà trouvées') and clarifying them in such a way that their proof is irresistible, thus putting theorems, once properly proven, on an epistemic par with axioms. Later, he calls proofs convincing (392), because their role is to convince us of their conclusions (416), something geometry does 'with irresistible force' (418). Additionally, proofs which conform to the precepts Pascal describes and finds in geometrical contexts cannot be met with any doubt (418). Of course, prior to their proofs being grasped, the *demonstranda* may be 'obscure', and indeed it behoves us to prove any statement that is initially even a little 'obscure' (419). On the basis of this evidence, we may ascribe *EP-Flow* to Pascal, indeed in its strong, highest-degree-transmission, guise. For as he puts it:

> Some of these [truths] may be drawn, by a necessary consequence, from common principles and established truths. Such truths one may be infallibly persuaded of; for, by showing the relationship they have to principles that have been granted, there is a necessity that cannot fail to convince. (415)

Given our deliberately broad characterisation of the EP in terms of the schematic relation *E* (see §2), Pascal clearly subscribes to its three core principles. However, the case for the subsidiary principles is much weaker. Pascal in the *Esprit* makes no pronouncements that settle *EP-Finite*, *EP-General*, or *EP-Independence*. There is a little more support available for *EP-Completeness*. Some passages, such as the following, may be read in such a way as to support the attribution:

> … when it [geometry] has arrived at the first known truths, it stops there and asks that they be granted, for lack of something more obvious from which to prove them: in such a way that everything that geometry puts forward is perfectly proved, either by the light of nature, or by proof. (400)

If we read 'everything that geometry puts forward' as 'any known/knowable/true geometrical sentence' then the respective form of completeness is implied. But if, more plausibly, 'everything that geometry puts forward' is read as 'any geometrical theorem', no commitment to a version of completeness follows: Pascal's point would simply be that all geometrical theorems have the stated property, which would be compatible with some geometrical truths, even some known ones, being neither theorems nor negations of theorems. More generally, there seems to be no unambiguous support for *EP-Completeness* in the *Esprit.*

In short, Pascal in the *Esprit* paints a picture of the Euclidean method that instantiates the core part of the EP.[76] Lakatos (1962: 10) believes that the EP has been 'on a great retreat' ever since. Although there is some truth to this,[77] the EP has in fact been influential until very recently. In a textbook first published in 1941, Tarski explicitly cites Pascal's conception of mathematical method as a prime influence on his own (1994: 109, n. 1). Despite being written long after the EP's apogee, we find Tarski articulating and defending a version of the EP which is, in rough outline, identical to Pascal's version. The method is designed to offer 'the highest possible degree of clarity and certainty' (1994: 109) and is explicitly based on the core Euclidean principles of *EP-Truth*, *EP-Self-Evidence*, and *EP-Flow* (1994: 110). Perhaps surprisingly, Tarski also writes that the *Elements* itself 'does not leave much to be desired from the standpoint of the methodological principles stated above' (1994: 111, n. 2). He even goes so far as to claim that the Euclidean method 'is the only essential feature which distinguishes the mathematical disciplines from all other sciences; not only is every mathematical discipline a [Euclidean] theory, but also, conversely, every [Euclidean] theory is a mathematical discipline' (1994: 112).[78]

Even more recently than this, Feferman describes the Euclidean view (again embodied by *EP-Truth*, *EP-Self-Evidence*, and *EP-Flow*) as being 'currently conventional' (1998: 77). To find such remarks in the work of leading mathematicians, logicians, and philosophers so long after the advances which made the decline of the EP possible, and so close to our own time, shows that Euclid's shadow is long indeed. Something like this view seems to be difficult to dislodge as the default view of mathematical method. It will be interesting, then, to compare the EP to a contemporary ideal of axiomatisation.

6 Descriptive Axiomatisation and the EP

Let us follow Detlefsen (2014: 61) in calling an axiomatisation designed to put logical structure on a body of facts or data a *Descriptive Axiomatisation.*

[76] There is also a strand in the *Esprit* which, in common with some of Descartes' remarks, could be construed as advocating *EP-Chauvinism* (see e.g. 417–8 or 425–6).

[77] An eighteenth-century example supporting Lakatos's contention is Alexis Clairaut's *Elements of Geometry* (1741), which explicitly declines to follow the Euclidean method in geometry; see especially Clairaut (1741: 10).

[78] Tarski's preferred term is actually 'deductive theory', but these are exactly the theories which he takes to accord with the EP's three core tenets.

These facts may, for example, be the truths of arithmetic, of geometry, or of some other area of mathematics; or they may be the truths of all mathematics; or of a non-mathematical science, such as physics or economics, or one or more of their subfields. Detlefsen (2014: 63) characterises Descriptive Axiomatisation in the following fourfold way[79]:

1. Axiomatisation generally takes place against the background of a *data set*.
2. The elements of this set are the commonly accepted sentences pertaining to a given subject area.
3. The basic purpose of axiomatisation is to organise a data set deductively.
4. To accomplish 3 fully, the axioms of a proposed axiomatisation must be *descriptively complete* – that is, all elements of the data set must be deducible from the axioms.

How does Descriptive Axiomatisation relate to the EP? Its first three tenets are not built into the EP, and one can imagine someone accepting the EP without endorsing any of them. A Euclidean foundationalist need not commit to there being a background data set (tenet 1), or to its elements being commonly accepted sentences (tenet 2), or to the purpose of axiomatisation being to organise the data (tenet 3). But although our characterisation of the EP does not rule out a Euclidean presentation *ab initio*, in practice the axioms are given for a branch of mathematics that is already at some stage of maturity, and the intention in Euclid's *Elements* and its successors is not merely to make mathematical discoveries by proving novel theorems, but to derive the commonly accepted sentences of some subject area. Given that background, a Euclidean will in practice quite naturally, although not necessarily, subscribe to something like the first three tenets of Descriptive Axiomatisation. (The claim that the organising function of axiomatisation is *the* basic function would presumably need to be weakened.) As for Descriptive Axiomatisation's fourth tenet, it is built into *EP-Completeness*: theorems follow from axioms, which together with the rules must be sufficient for inferring all relevant theorems.

Unlike proponents of Descriptive Axiomatisation, Euclideans insist on the epistemological significance of the organisational structure in terms of the flow of some epistemic good. This is a central difference between the EP and Descriptive Axiomatisation. According to the EP's foundationalist perspective, the epistemic relation we bear to the axioms is primary, and our knowledge of the axioms explains our knowledge of the theorems, or, at the very least, inferring theorems from the

[79] Detlefsen understands 'a given subject area' in the second tenet as locally restricted, that is, as applying to a subfield of mathematics rather than all of mathematics. In contrast, we allow both a global as well as a local interpretation. The italics in the first and fourth tenets are Detlefsen's; we have anglicised his spelling, for consistency with our own.

axioms improves the epistemic standing of the former. Descriptive Axiomatisation contains no analogue of this principle; it is quite consistent with Descriptive Axiomatisation that our epistemic relationship to the *data* is primary, and that our knowledge of the axioms is owed to the fact that the data are deducible from them. We'll return to this issue in §9, but for now it's worth noting that in a Euclidean theory the derived theorems do not perform an epistemological function analogous to the role of data in the observational sciences.

More generally, compared to the EP, Descriptive Axiomatisation is relatively light on philosophy. It describes an ideal that a logician or mathematician might sign up to without incurring many epistemologically controversial commitments. Indeed, we find E. Huntington recommending it in the following terms in a specifically geometric context:

> [a] miscellaneous collection of facts about angles does not constitute a *science*. In order to reduce it to a science, the first step is to do what *Euclid* did in geometry, namely, *to select a small number of the given facts as axioms, and then to show that all other facts can be deduced from these axioms by the methods of formal logic*. (1911: 158)

Detlefsen, who also cites this passage (2014: 63), comments that, suitably generalised, Huntington's remarks expressed a common view of the aims and purposes of local (subfield-specific) axiomatisation in the nineteenth and twentieth centuries. We find some version of it in different writers, though not all adopt Huntington's very strong completeness requirement. To cite but one example, Arend Heyting, Brouwer's influential disciple, characterises the axiomatic method as follows in the introduction to his book on projective geometry (1980: 5)[80]:

1. A complete list of the fundamental notions of the theory is given.
2. Every other notion is reduced to the fundamental notions by explicit definition. These definitions must be of such a nature, that everywhere, except in the definition itself, the definiens can be substituted for the definiendum. Consequently, we could in principle dispense with the defined notions.
3. A complete list of fundamental theorems (called axioms) is given.
4. Every other theorem is deduced from the axioms by logical reasoning.

Incidentally, Heyting comments that 'an axiom is no longer regarded as an indubitable truth' (1980: 1), thereby clearly distinguishing his conception of Descriptive Axiomatisation from the EP. But despite the decline of the EP after the seventeenth century, a contemporary ideal of axiomatisation inspired by the *Elements* survived until at least the latter years of the twentieth century.

[80] Curiously, this famous intuitionist chose to use classical logic in this particular book (1980: 5).

Although philosophically lightweight by comparison, the tenets of Descriptive Axiomatisation would be largely congenial to advocates of the EP, demonstrating the lasting, if waning, influence of this way of thinking about mathematics.

7 The EP Assessed: Core Principles

The first half of this essay is now complete. In it, we reconstructed the Euclidean Programme and compared it to some historical sources. It is now high time to assess it. In doing so, and to keep to this Element's word limit, we shall have to restrict our focus and be less expansive than we might have wished to be in several ways.

First, we assume some logic and set theory, of the sort covered in introductory texts. Second, we note that the EP is a form of epistemological foundationalism in the domain of mathematics. It would be illuminating to place the EP in a more general epistemological context by considering which criticisms of it in §§7–8 are specific to the EP and which generic criticisms of epistemological foundationalism, but that would take us beyond our remit and word limit. Third, there is reason to think that for several important historical figures, the EP or something like it was understood as a rational or methodological ideal rather than a theory of *actual* mathematical knowledge, justification or rational belief. Criticisms of the EP understood as a descriptive thesis might therefore seem irrelevant to its status as a normative thesis or methodological ideal. Unfortunately, there is no room to examine whether the EP should be a methodological ideal or not. We can, though, say something in defence of our approach. (i) Some of our criticisms apply to the EP understood even as a normative thesis. (ii) Methodological ideals in epistemology have nevertheless been aimed at characterising the highest epistemic state that creatures like us can achieve. As Pasnau puts it, '[t]he objective is not to identify standards that only a god could achieve' (2017: 7). It would be a major blow to the EP if various fields in mathematics do not fit its mould and cannot obviously be improved by being made to fit it. (iii) We follow a broad naturalist trend in thinking that the philosophy of mathematics should try to make sense of mathematical epistemology rather than dictate to it. If there is a significant discrepancy between an ideal Euclidean picture on the one hand, and the actual practice of mathematicians on the other, then either the ideal is wrong, or there is something seriously amiss with the practice. Without deferring uncritically to the standards exhibited in modern mathematics, we should nonetheless give practice the priority in the resolution of such disputes, in the absence of particularly strong reasons to the contrary.

Finally, as we cannot survey all mathematics, we limit ourselves to three areas. The first is set theory, the second arithmetic, and the third, a representative from a broad class, group theory. We shall also say a word or two about a few other areas, where relevant.

Our three representatives were chosen for the following reasons. First, we wanted to err on the side of charity. All three are fields of mathematics that are at first sight favourable to the EP, especially when compared to other areas like fluid dynamics or complex analysis. Our criticisms of the EP will be all the more powerful if we show it does not succeed in these areas. Arithmetic especially is chosen because of its centrality to the subject and because, as we shall see, it appears to be the 'specific' theory (defined shortly) of mathematics most closely exemplifying the EP. Set theory earns its place because it is generally regarded as a foundation of mathematics, in the sense that all standard mathematical structures can be represented as set-theoretic structures, and any standard mathematical claim can be expressed in set-theoretic terms. The EP is linked to a foundational ideal, so it is very natural to consider whether it applies to the best contemporary candidate to play this role – set theory. As a preamble to justifying the third, we observe that some mathematical theories are typically thought to be about individual structures: arithmetic, real analysis, complex analysis, and Euclidean planar geometry are all examples. Other mathematical theories, such as group theory, field theory, and topology are instead thought to apply to a host of structures. The former are thought to have an intended interpretation – arithmetic is specifically about *the* natural numbers, for instance – and the latter are thought to be about *any* structure that satisfies their axioms. We accordingly call the former 'specific' and the latter 'general'.[81] We choose group theory as the third in our trio of theories as it is an unambiguous example of a general theory. In our assessment of the EP, it does duty for all such theories.

Where exactly the line between specific and general theories falls is a disputed matter; for example, it is not clear whether set theory, often thought of as specific, is instead general. Arguably, at least one mathematical theory

[81] Feferman (2000: 403) calls the former 'foundational' and the latter 'structural'; Shapiro, following a suggestion of Hellman's, calls them 'assertory' and 'algebraic' respectively (2009: 177). 'Foundational' is far from ideal because it suggests that a theory with an intended interpretation serves as a foundation for mathematics. The pair 'assertory' and 'algebraic' are unsatisfactory because 'algebraic' theories consist of assertible sentences (see the discussion to follow in the main text), and many theories outside algebra are 'algebraic'. Of course, general theories as we call them still have a definite subject matter in some sense. For example, it is generally thought that group theory concerns a class of set-theoretical structures (those satisfying the group axioms), and set-theoretical facts can be made use of in proving group-theoretic results (e.g. that two groups are isomorphic). But there isn't a specific such structure to which the axioms are intended to apply.

must be specific. This is the theory, which Hilbert called metamathematics,[82] that allows for mathematical investigation of mathematics taken as a whole. Today, logicians who investigate mathematics as a whole do so within an (often implicit) set theory, instead of Hilbert's (weaker) preferred metatheory of finitist mathematics. This reinforces the case for taking set theory to be specific rather than general, and in the following sections that will be our assumption.

7.1 EP-Truth

This first core principle is relatively straightforward to assess. Any axiomatisation that purports to represent facts about a particular structure – for example, Peano Arithmetic about the natural numbers – ought to consist of true axioms. So, specific theories should satisfy *EP-Truth*. This applies to both arithmetic and set theory.

What about general theories, such as group theory? These theories' axioms are stipulative of their subject matter. The group axioms – the closure of the group under its binary operation, this operation's associativity, the properties of the identity element, and the existence and properties of inverses – define what it is to be a group. Any suspicion of the group axioms being incorrect would be misplaced, since they are by definition correct about groups. For instance, any structure equipped with a binary relation under which some element lacks an inverse is definitionally excluded from being a group and is at best a monoid. It is, of course, a further question whether any structure of the type axiomatised by a general theory exists. In short, *EP-Truth* holds for general as well as specific theories; but in the former case, the axioms are stipulatively true, whereas in the latter, truth is understood in the usual, non-stipulative way.

Naturally, there is more to say about what stipulative truth comes to. Shapiro (2009: 177) claims that because the axioms of a general theory (in our terminology) are definitions they are not asserted, and thus that we do not know them. But it seems to us that at least some stipulations can be asserted and known. Imagine an emperor who has been granted authority by all the assembled senators to nominate the new capital of the empire. Using a declarative sentence, he stipulates that it is to be Ravenna rather than Rome. His statement is an assertion, subsequently known by all present. Similarly, that triangles have three sides seems obviously true, and to be common knowledge, even if true by stipulation. Moreover, Shapiro's account entails that the theorems of general

[82] Which, for Hilbert, was the one and only mathematical theory which conformed to the EP (see Hilbert 1925), its arguments proceeding from self-evident intuitive axioms via self-evidently and intuitively truth-preserving rules. Lakatos opines that Gödel's second theorem 'was a decisive blow to this hope for a Euclidean meta-mathematics' (1962: 20). For how a Hilbert-inspired instrumentalist might react to this theorem, see Paseau (2011).

theories are themselves not known, since they are inferred from stipulated axioms (2009: 177). But we cannot see any good reason to think that elementary consequences of the group axioms, such as the uniqueness of the identity element in any group, are unknown. Rather, they seem to be the sort of thing that is known by almost anybody who knows what a group is. There is good reason, then, to think that the group axioms are stipulated truths about groups.

As for theorems, we note that they must be implied by axioms via a sound system of rules. Since axioms are true (perhaps stipulatively), it follows that theorems must be true as well. This requirement is very much still in place today, for both specific and general theories – and in particular for set theory, arithmetic, and group theory. If we infer a conclusion from premises all true about a collection of structures, we want the conclusion to be true about those same structures as well. Likewise for a conclusion about a particular structure-type inferred from premises true of that same structure-type.

7.2 EP-Self-Evidence

It is very hard to assess *EP-Self-Evidence* in its abstract formulation; to each 'epistemic good' there corresponds a different version and a corresponding set of reasons to accept or reject it. So, we suggest understanding the principle as it would generally be understood today, in terms of justification. Thus understood, it states that all axioms are graspable and self-evident, and that if a subject clearly grasps a self-evident proposition then she is justified in believing it to the maximal degree.

To assess this justificatory version of *EP-Self-Evidence*, we distinguish two senses of self-evidence. A claim may be self-evident if it is maximally obvious; call this *maxevident*. Or it may be self-evident if its evidence is owed to itself only – it is evident by itself and neither receives nor requires outside support – which we may call *ipsevident*.[83] These two senses are recognised by the dictionary[84] and should be sharply distinguished but are often conflated. The principles linking maxevidence and ipsevidence to the notion of justification are roughly as follows[85]:

[83] From the Latin *ipse* meaning 'itself'.

[84] Here is the OED entry (December 2019 ed.) for the adjective 'self-evident': 'Evident by itself; requiring no proof or explanation; obvious, axiomatic.'

[85] The elucidations have a suppressed time index t. For the definition of ipsevidence, compare Jeshion (2001: 953). There is a natural question to raise at this point, namely, whether *EP-Self-Evidence* implies *EP-Truth*. That is, must ipsevident or maxevident principles be true? The answer depends on your conception of justification. On a highly internalist conception of justification, a subject may be maximally justified in believing p without p being true; but on a more externalist conception, this may not be possible.

p is maxevident ~ If a subject clearly grasps *p* then her belief in *p* is justified to the highest degree.

p is ipsevident ~ A subject can have no more justification in believing *p* than she acquires by clearly grasping *p*.

We intend these not as strict definitions, analyses, or explications, but as elucidations. Clearly, more could be said about the right-hand terms (e.g. what is involved in clearly grasping *p*?), but there is one ambiguity that requires immediate attention. In the first elucidation, does being justified to the highest degree in believing *p* mean being justified to the highest degree one can be in believing *any* proposition, or does it mean being justified to the highest degree one can be in believing *p* specifically? (Perhaps you are as justified in believing p_1 as much as it is possible to be, but less justified than you could be in believing p_2.) Presumably the former, otherwise the EP risks selling axiomatisations short. For if axioms are merely maxevident in the latter sense, one may not be very justified in believing them just by clearly grasping them. Geometry, for instance, could then be a house of cards, built from frail yet maxevident axioms. Similarly, mere ipsevidence will not do: a clearly grasped axiom for which we have very little justification and can obtain no more would qualify as ipsevident but would not serve the EP's foundational purposes.[86]

Now, there are several theories of maxevidence and ipsevidence on the market. A rationalist might insist that the evidence of such propositions is based on mathematical intuition, whereas an empiricist might see it as based on the meanings of the words. There is no need to broach any such accounts of the source of self-evidence for our purposes. Turning first to set theory, even a brief glance at the contemporary axioms is enough to see that (the justificatory version of) *EP-Self-Evidence* does not generally hold.

Following previous authors, notably Russell, Gödel, and Maddy, we may distinguish *intrinsic* from *extrinsic* evidence in set theory.[87] This distinction is admittedly vague; more problematically, it is drawn in different places by different authors. But it remains serviceable, and our use of it should be uncontroversial. Our preferred construal is to take extrinsic evidence for a principle to consist in its instrumental value, in drawing consequences,

[86] A third elucidation is possible: if a subject clearly grasps *p* then she is justified to the highest degree possible *for a proposition with the subject matter of p*. One might want to distinguish this sense if one thinks, for instance, that the axioms of arithmetic are inevitably more evident than the axioms of set theory, because arithmetic's subject matter is more readily apprehended. Note, however, that this sense of evidence will not serve the Euclidean's purpose, for broadly the same reason that ipsevidence will not.

[87] The intrinsic/extrinsic distinction, of course, applies more generally. We return to it in §9.

forging connections between different areas, making for better explanations of the relevant data and the like. This is the kind of evidence on which theoretical principles in science are mostly, if not exclusively, based. Intrinsic evidence we may take to be non-extrinsic evidence: the sheer obviousness or plausibility of a principle, as well as how it fits with the conception of the subject matter.[88]

However we distinguish intrinsic from extrinsic, it is now almost universally appreciated that at least *some* of the justification for the standard set-theoretic axioms is extrinsic. For example, the Axiom of Infinity, which states that an inductive set exists, has some intrinsic plausibility on the most common (iterative) conception of set, but is also required for mathematics beyond the finitary. The Axiom Scheme of Replacement states (schematically) that if a relation functionally maps the elements of a set to some other elements, then those other elements also form a set. Again, we may debate how much intrinsic evidence this principle possesses, but its inclusion amongst the axioms is in no small part owed to its fruitful and necessary consequences.[89] Axioms such as Infinity or instances of Replacement are quite different in this regard from some others that wear their plausibility on their sleeve. This point is familiar enough from the philosophical literature on set theory that we will not dwell on it. Some of the more detailed arguments may be found in Maddy (1988), which, more than three decades on from its publication, remains an excellent account of the justification behind set theory's axioms and axiom candidates.

What is the upshot for our discussion of (the justificatory version of) *EP-Self-Evidence*? Today's set-theoretic axioms are usually supported by a variety of evidence, some intrinsic, some extrinsic. It would be quite a stretch to suppose that *all* these axioms are self-evident in anything like the historically prevalent sense. They are neither ipsevident nor (*a fortiori*) maxevident.

One might suspect, however, that *EP-Self-Evidence* should indeed apply to our foundational theory, but that this theory is not set theory. It is a rival foundation, the leading candidate being category theory. We have two things to say in response. First, our focus on the axioms of set theory so far does not strictly speaking presuppose set-theoretic foundationalism – the idea that in some sense set theory is a foundation for mathematics. Set theory is an example

[88] We make no claim that intrinsic and extrinsic considerations are completely independent from one another. For example, a principle may more evidently give a correct characterisation of its subject matter when one realises that it has some particularly elementary theorems as consequences. Thanks to Neil Barton for this point.

[89] For example, without Replacement, one cannot form the ordinal $\omega + \omega$ and hence neither the sets at or beyond that stage in the iterative hierarchy. Famously, Friedman (1971) showed that Replacement is essential for the proof of Borel Determinacy.

of a contemporary axiomatised area of mathematics that conflicts with *EP-Self-Evidence*, whether or not it serves as a mathematical foundation. The EP was held up as a paradigm for *all* mathematical theories and even for non-mathematical theories in a variety of disciplines. Withholding its application to a significant area of mathematics, set theory, be it foundational or not, is a significant retreat from how the EP was historically conceived.

Second, analogues of the points just made about set theory equally apply to a category-theoretic foundation. We cannot properly show this here, but the point is not controversial: the justificatory version of *EP-Self-Evidence* no more applies to a category-theoretic foundation than it does to the standard set theory, Zermelo–Fraenkel-Choice (ZFC). It's very hard to see, for example, how the axiom positing a category-theoretic version of the natural numbers (a natural numbers object), usually found in such a foundation, could be any more intrinsically justified than set theory's Axiom of Infinity. Also, a category-theoretic version of set theory's Axiom Scheme of Replacement must be added to any such foundation to render it as strong as ZFC, and this axiom is also far from self-evident. Its justification heavily rests on its mathematical consequences.

What about arithmetic? With no upheaval in the history of the discipline to rival those that have beset geometry and foundations, this might look like promising ground for the EP. So, are the axioms of, say, Peano Arithmetic self-evident? For clarity, let's have a list of the axioms. The first-order version of the theory, PA^1, has as non-logical symbols the constant symbol 0, the one-place function symbol S, and the two-place function symbols $+$ and $\times$. PA^1's axioms include the six statements

$$\neg \exists x(0 = Sx)$$

$$\forall x \forall y(Sx = Sy \rightarrow x = y)$$

$$\forall x(x + 0 = x)$$

$$\forall x \forall y(x + Sy = S(x + y))$$

$$\forall x(x \times 0 = 0)$$

$$\forall x \forall y(x \times Sy = x \times y + x),$$

as well as the infinitely many instances of the induction scheme

$$[\phi(0) \wedge \forall x(\phi(x) \rightarrow \phi(Sx))] \rightarrow \forall x \phi(x).$$

The second-order version, PA^2, can be axiomatised by three axioms, the first two of which are shared with PA^1 and the third of which generalises the induction scheme:

$$\forall P([P(0) \wedge \forall x(P(x) \rightarrow P(Sx))] \rightarrow \forall x P(x)),$$

which wraps the infinitely many instances of the first-order induction scheme, and many more, into a single principle. In this theory, addition and multiplication are explicitly definable, making the remaining axioms of PA^1 redundant.

The two axioms common to PA^1 and PA^2, and PA^1's four further non-induction axioms, seem to us evident enough to qualify as self-evident. But the induction principle seems to fall short of self-evidence. The first point, which applies to PA^1, is that all but the smallest instances of the first-order induction axiom are too large to be graspable by human beings, though they are graspable by mildly idealised versions of us (see §7.3). The second point is that the induction axiom, in schematic or axiomatic form, seems to fall short of the epistemological standing of the other axioms, suggesting that it fails to be self-evident. To summarise a by-now infamous episode, the mathematician Ed Nelson claimed at one point that PA^1 was inconsistent and believed he had a proof of this fact. Nelson later retracted his claim. If Nelson's putative proof had at least some *prima facie* plausibility, we would presumably have looked to an instance of induction as the culprit rather than to a successor axiom or an addition or multiplication axiom. Although the possibility is far-fetched, it highlights the fact that our allegiance to the axioms of PA^1 is not monolithic: the first six axioms are *more* evident than at least some instances of induction, which implies that the latter cannot be quite maxevident.

One could reply that although induction (either in its second-order axiomatic or its first-order schematic form) *is* self-evident, the fact that it is self-evident is not itself self-evident. In other words, perhaps Nelson was confused: induction is self-evident, to him as to anyone else, but he mistakenly thought it wasn't. But this seems to be the wrong diagnosis. Nelson, an expert number theorist, conceived of the natural numbers as constructions, rather than a pre-existing totality, hence rejected impredicative instances of the induction scheme (1986: 1). He held on to this belief even after retracting his claim that PA^1 was inconsistent. Many other first-rate mathematicians, including Henri Poincaré, have rejected instances of induction on predicativist grounds. Nelson and other predicativists have thought hard about induction – indeed, harder than most, since they appreciate that in its full generality it has impredicative instances. They have clearly grasped induction but still reject it. What this shows is that the induction principle is not quite as evident as the other axioms. The way we would put it is that the induction principle is evident on the classical conception of the natural numbers, but not on the predicativist conception. We return to this notion of evidence on a conception below in §9.1.

Finally, what of a general theory such as group theory? Are its axioms self-evident? They seem to be. After all, a subject who clearly grasps the meaning of the term 'group' will be maximally justified in her belief that all group operations are associative. Of course, the axioms of a general theory are stipulated, so they are not self-evident in an interesting sense. It also may be far from self-evident whether the axioms of a general theory describe even a single structure. But such axioms nonetheless fit the bill, because they are maxevident: once a subject grasps the axiom that for any group, the group operation is associative over its elements, she has the highest possible degree of justification for that axiom.

The fact that the axioms of group theory are self-evident does not mean that their application is. Suppose that the axioms of group theory supplemented with some hypotheses about the kinds of groups we are interested in (e.g. finite and abelian) entail some theorem p. We might come across a structure S and wonder whether S satisfies the supplemented set of principles (that apply to all and only finite abelian groups) and therefore wonder whether p holds in S. It may be an entirely non-trivial enterprise to establish that S satisfies the principles and thereby that p is true of S. In practice, many problems involving groups are of this kind: not pure deduction from axioms, but working out whether a given structure is a group and if so of what kind, so that established theory may be applied to it. Moreover, beyond the first few pages of a textbook exposition, group theory uses results from many other parts of mathematics. The existence of homomorphisms between various structures, for example, does not follow from the group-theoretic axioms, but from the ambient set-theoretic world in which groups live. Substantive progress in group theory, beyond first baby steps, relies on much more than the group axioms and logic.

In sum, the axioms of group theory may be self-evident, but this fact is not terribly important. Applying the axioms is at least as important, and work on groups goes far beyond deduction from group-theoretic first principles (i.e. fundamental axioms plus supplements).

We must at this point issue a caveat about the understanding of 'maxevident'. It may be rational to withhold a small degree of credence in any principle, however plausible it may strike us. This may be for the sorts of reasons Descartes elaborated in the *Meditations* – can we rule out that an evil demon is deceiving us? – or others. Investigating the nature and size of this 'Cartesian shortfall', if we may call it that, would take us too far into general epistemology and away from the Euclidean Programme. We note simply that if such a Cartesian shortfall exists, any self-evident axiom satisfies *EP-Self-Evidence* only *modulo* this shortfall. More generously, we can say that such an axiom is maxevident, as long as we understand 'highest degree' so as to allow for the existence of this shortfall.[90]

What is the upshot? *EP-Self-Evidence* does not apply to many of the axioms of contemporary set theory and doesn't even apply to all the axioms of arithmetic. The axioms of group theory all seem self-evident, however. In short, although the notion of self-evidence does apply to a number of axioms in diverse areas of mathematics, we cannot unreservedly say that *EP-Self-Evidence* applies to contemporary mathematics as a whole.

7.3 EP-Flow

As with *EP-Self-Evidence*, we may take the *E*-relation in *EP-Flow* to be justification. Discussion of *EP-Flow* can largely proceed in a unified manner, as it does not depend on whether the domain is set theory, arithmetic or group theory, or any other part of mathematics for that matter. However, one point does apply to specific rather than general theories: justification in a specific mathematical theory is *two*-directional. It flows 'top-down', from axioms with some intrinsic plausibility to theorems, but also – and perhaps even more importantly – 'bottom-up', from established claims to axioms that organise them. The axioms of arithmetic, for example, gain at least part of their justification from the fact that they imply elementary claims such as '$2 + 2 = 4$' or '5 and 7 are twin primes'. This is one reason why we think the EP does not fully apply even in perhaps its most promising domain, that of arithmetic. It is precisely to this general point that Lakatos alludes in his characterisation of the so-called Empiricist Programme, as we shall see in §9.

A related point is that finding a second (or third or fourth) proof of some mathematical fact increases our confidence in it. This impugns the strong version of *EP-Flow*, which in combination with *EP-Self-Evidence* implies that a mathematician who grasps a single proof of p is maximally justified in believing p. If grasping a second proof of p increases a mathematician's justification for believing p, they could not have been maximally justified previously.

There is another problem in the vicinity for *EP-Flow*. The version of *EP-Flow* in which the relation E is taken to be (some species of) justification reads as follows: if a conclusion follows from some premises, and the subject clearly grasps this, and has a high degree of justification in the premises, she thereby

[90] Here's a very simple – indeed, simplistic – model of what this might involve. Suppose the 'Cartesian shortfall' is modelled probabilistically as 0.005, meaning that it is not rational to believe any claim with credence more than 0.995. In that case, a maxevident principle would be any rationally held with degree of belief 0.995.

has a similarly high degree of justification in the conclusion. This version of the principle presupposes that justification is a graded notion, surely an uncontentious claim. At the start of the season, Luke may be justified in believing that Arsenal will not win the Premier League (based on the team's form the previous season, their summer signings, etc.), but he will be *more* justified in that belief after seeing them lose their first three games on the trot.[91] The justification version of *EP-Flow* asserts roughly that there is little loss of justification in inferring a theorem from some axioms. In its strongest version, it asserts that there is *no* such loss of justification – no erosion whatsoever.

However, the strong justification version of *EP-Flow* is false if we accept a reasonable-sounding principle linking justification and rational credence. Suppose you have deduced q from p and that your rational credence in q is less than your rational credence in p; then your justification for q is less than your justification for p.[92] The principle focuses on the one-premise case for simplicity but is easily generalised. It seems fairly robust to counterexamples, but in any case we only need its application in the particular cases of interest. To show that the strong version of *EP-Flow* is false, it therefore suffices to show that inferring a conclusion from a premise can be rational-credence eroding.[93]

It is fairly clear that even a deductive argument need not, and typically will not, be rational-credence preserving. (Similarly for a non-deductive argument.) A reflective subject, aware of her limitations of reasoning power, memory, attention span, etc., who follows a deductive argument to its conclusion should typically give it lower credence than she gives the conjunction of the premises. For example, if the subject knows that in any ten-minute period there is a 1 per cent chance she will let an error slip, the credence she should give to a conclusion of an argument she has been working through for a couple of hours and whose premises she collectively believes to degree 0.9 should be considerably less than 0.9 (if that is her only evidence for the conclusion).[94]

A point along these lines is reasonably well-established in epistemology. Schechter (2013), for example, argues in detail that justification is not always preserved by deductive reasoning by thinking about 'long sequence arguments' – arguments with a huge number of steps. Rational degree of

[91] That is not to say that a subject's justification for belief p is always comparable in strength with their justification for belief q, for any p and q.

[92] A rational credence (or degree of belief) should not be confused with a credence *simpliciter*: the latter is a purely subjective notion, whereas the former is not. This section more generally assumes some facts about rational credences without invoking a full-fledged account such as Bayesianism.

[93] The remainder of this section adapts some ideas in section 5.4 of Paseau (2016).

[94] Jaffe cites Serre as claiming that the perfect mathematical paper is a myth (1997: 135), for something like these reasons. We are imperfect logicians and mathematicians, so even the best work in a complex area will contain numerous small gaps and minor mistakes.

belief drops ever so slightly with each deductive step, in a way that aggregates. Even if not every step erodes our justification, a long sequence will.

All this may be granted, but the reply might be that deductive arguments are rational-credence preserving for *ideal* subjects. Our response to this is twofold. First, as stated at the beginning of §7, we are less interested in the epistemology of ideal beings than in the epistemology of actual ones – flesh-and-blood users of mathematics. Our question is whether mathematical practice lives up to the standards of the EP, not whether idealised practice does so. And actual mathematicians are very much subject to credence and justification erosion.

Second, the notion of an ideal agent can be made precise in a variety of ways. There is no doubt a legitimate sense of 'ideal agent' in which such a subject is logically omniscient,[95] and perhaps for them deductively valid arguments are rational-credence preserving. And indeed, it seems as if the Euclidean Programme itself is designed for agents which are idealised to some degree and in some ways. Even Euclid's geometer can draw line segments of arbitrary length, which no actual agent is capable of. More generally, requirements like *EP-Completeness* demand only that certain propositions are deducible from the axioms; but a deduction can be so long and complicated that no actual mathematician could produce it before expiring. But in general, the EP understood as a methodological ideal seems to operate with the standard idealisation which is, according to Shapiro, 'invoked throughout mathematics, where we ignore limitations on attention span, memory, lifetime, and the like' (2009: 194). In other words, idealised agents are mathematically of a kind with ordinary human beings: they can just do more of the same but better and are not subject to our practical limitations.[96]

Ideal thinkers in the just-specified sense are, however, still subject to inferential uncertainty. Suppose that an idealised agent can be rationally certain that they have correctly applied an inference rule when they have and can always spot mistakes when they have not. Still, this sort of ideal subject might well give credence less than 1 to the proposition that some deductive rule is truth-preserving; they might not be *philosophically* infallible about whether a given notion of logical consequence is correct. Consider Double Negation Elimination: the inference of p from $\neg\neg p$. This classical rule is questioned by intuitionists and intuitionist sympathisers, a camp that includes L. E. J. Brouwer, Arend Heyting, and Michael Dummett – formidable thinkers all. Even a committed classical logician, given the history of this dispute, would be wise to lower their credence in the rule's truth-preservation a smidgen,

[95] See, for example, Clarke-Doane's notion of a *cognitively flawless* agent (2013: 471).
[96] See Wrigley (2022b, §2) for further discussion of this issue.

to think it is less likely to be truth-preserving than, say, either of the Conjunction Elimination rules.[97] For is it epistemically impossible that, say, Dummett's case for intuitionism is correct?[98] We suggest not: it seems rationally permissible to suppose that Dummett's arguments are sound and that they imply that Double Negation Elimination is not logically valid. A similar point, with varying degrees of force, can be made about modus ponens for the natural-language 'if … then' construction,[99] or the claim that anything follows from a contradiction, challenged by dialetheists, and so on.

Mathematical enhancement should therefore be distinguished from philosophical idealisation. The former falls very short of what would be required, for example, to provide a rationally unimpeachable refutation of Dummett's philosophical position. Though progress in philosophy is certainly possible, we cannot reach rationally unshakeable philosophical conclusions merely through enhancing our existing mathematical powers. So, as far as the strong justification version of *EP-Flow* is concerned, the conclusion stands. When mathematically enhanced mathematicians perform an inference, especially a long one, their rational credences may be eroded. And their justification for a conclusion of a deductively valid inference may be less than their justification for its premise(s). *A fortiori*, this is true of flesh-and-blood mathematicians, who are also subject to limitations which their mathematically enhanced counterparts are free from.

Similar considerations broadly apply to the weaker version of *EP-Flow* stated in §2 with relation E instantiated by (some species of) justification. Rational credence may be eroded by inference for all but philosophically infallible idealised subjects, so that someone with very high credence in the axioms may end up, quite rationally, with a not-particularly-high credence in a theorem derived from a high-credence conjunction of axioms, especially if the derivation is long. Hence *EP-Flow*, understood as applying to justification, is not tenable in either its strong or weaker version (given the link between justification and rational credence). This moral applies to all the branches of mathematics.

We conclude that only one of the EP's three core principles, namely, *EP-Truth*, is unambiguously left standing as far as modern mathematics is concerned.

8 The EP Assessed: Subsidiary Principles

§7 covered the three core EP principles. We now assess the EP's four subsidiary principles.

[97] That is, from A & B, infer A; and from A & B, infer B. [98] Dummett (1973).

[99] McGee (1985); Lycan (2001: ch. 3).

8.1 EP-Finite

We remarked in §2 that a theory satisfies *EP-Finite* if it has a finite presentation, that is, the number of its non-schematic axioms plus the number of its schemes is finite. This finitude requirement is reasonable on the assumption that our minds and lifespans are finite, so that any axiomatisation which human mathematicians can give or understand must be finitely describable. Even the idealised mathematician, who is arguably the subject of the EP, can at no time have comprehended infinitely many axioms. But a theory which includes infinitely many axioms simply by virtue of including finitely many schemes is not problematic in a way that is epistemologically relevant.

The finite-presentation requirement covers the overwhelming majority of (formalisations of) theories used by contemporary mathematicians. It covers, for example, ZFC set theory, which contains two schemes with infinitely many instances[100] and a finite number of other axioms. The same goes for extensions of ZFC, ranging from widely accepted weak extensions (e.g. ZFC plus the claim that there exists an inaccessible cardinal) to much stronger ones investigated by set theorists. The same goes for the standard theories of arithmetic, analysis, groups, etc. Recalling the point briefly made in §2, we also note that all these axiomatisations in fact involve a manageably small number of axioms. Naturally, contemporary logicians investigate a very wide array of theories, some of which do not admit a finite presentation.[101] But the latter are not typically (or ever, so far as we know) theories that could be said to represent mathematical reasoning about a particular domain.

When it comes to rules of inference, contemporary logical systems usually consist of finitely many of them. These systems may be Hilbert-style, with say, just modus ponens as the propositional rule and a small (finite) number of rules pertaining to the universal quantifier. Natural-deduction systems are another popular approach. In a typical natural-deduction system, there are at most two introduction and at most two elimination rules per connective, and thus finitely many rules, since the system contains only finitely many connectives.

[100] Namely, the Axiom Schemes of Separation and Replacement (though the former's instances are derivable from the other axioms, as we shall observe later).

[101] For example, to demonstrate that the statement of the upward Löwenheim–Skolem theorem depends on the size of the language, one may consider a first-order theory with κ inequations of the form $c_i < c_j$ with $i < j < \kappa$ and κ an infinite cardinal. Another example is True Arithmetic, which consists of all the true sentences in the language of arithmetic.

As the evidence stands, then, *EP-Finite* holds good. For it to fail, we would need to observe mathematicians making use of a *truly* infinitary theory, which included infinitely many axioms that are not instances of a scheme, or else infinitely many schemes infinitely many primitive rules, or rules that were infinitary in nature. But such theories are not mainstream fare, even in contemporary mathematics.

8.2 EP-General

Axiomatisations usually conform to *EP-General* because their axioms either are or can be recast as general propositions. Some examples: the only specific entity mentioned in usual axiomatisations of set theory is the empty set, definable as the only set with no members; in arithmetic, 0 may be defined as the only element whose sum when added to itself is itself; in group theory, the identity element is the only element e such that $e + x = x = x + e$ for any element x of the group (where + is the group operation); in the theory of real fields, 0 may be defined in the same way as the 0 for arithmetic, and 1 as the only element other than 0 whose product with itself is itself; and so on.

Now, of course, nothing prevents us from adding undefined constants to a theory, and model theorists investigate all sorts of axiomatic theories containing undefined terms. To pick just one example at random from a bottomless bag, one may be interested in the properties of the theory of dense linear orders without endpoints with two constants c_1 and c_2 such that $c_1 < c_2$ but in which no other information about c_1 or c_2 is given, so that neither of these is definable. In the light of that, our conclusion is that in standard mathematical practice, *EP-General* applies to the great majority of theories that arise 'naturally', as long as we allow particular entities to feature if definable in general terms. It certainly applies to our chosen trio of set theory, arithmetic, and group theory.

One could go further and ask *why EP-General* applies so generally, and why in particular it seems to apply universally to specific theories (which have an intended interpretation). Doing so, however, would take us too far into the metaphysics of mathematics. We simply note in passing that any satisfactory metaphysics of mathematics should be compatible with this fact. Stewart Shapiro's *ante rem* structuralism, for example, whatever its other defects, is well-placed to explain why *EP-General* holds, precisely because it sees mathematical entities as having only 'structural' properties.[102] On that view, any entities that can be picked out at all can be picked out in 'structural', and so, presumably, general terms.

[102] Shapiro (1997).

8.3 EP-Independence

While information about the mutual independence, or lack thereof, of axioms in a given mathematical theory is no doubt of mathematical value, modern mathematics does not insist on *EP-Independence* in any strict sense.

We first give an example to show that axioms of set theory are not independent. There is a finite set Φ of instances of ZF set theory's Separation Scheme such that Φ, together with the axioms of ZF other than the instances of Separation, proves all the instances of the Separation Scheme.[103] So why include all the instances of Separation? The system Ax_1 = ZF is preferable to Ax_2 = ZF without the Separation Scheme + the finitely many instances of Separation in Φ, because Ax_1 is simpler, axiomatically speaking, than Ax_2. A single simple scheme, with infinitely many instances, is preferable to a finite miscellany, despite the fact that infinitely many instances of the scheme are logically redundant. In other words, modern set theory violates *EP-Independence* out of a preference for economy of axiomatic presentation.

The standard axiomatisation of arithmetic also contains some redundancy. For example, some instances of Peano Arithmetic's induction axiom imply others, some are logical truths (so need not be posited as separate axioms), and so on.

To this, the reply might be that *EP-Independence* should be assessed at the level of schemes.[104] The idea is that a scheme should not be identified with its instances, but should be thought of as a single thing that entails or unifies its instances. A scheme would then be considered independent of the rest of a theory if it is not the case that *all* its instances can be derived from the remainder of the theory. On that understanding, does *EP-Independence* hold? It does for PA^1, and also for PA^2, so long as the multiplication and addition axioms are omitted, being derivable from the axioms for successor and induction. However, we observe that in several standard presentations of PA^2,[105] these axioms are retained for their convenience and their usefulness in facilitating a comparison to weaker systems, including PA^1.

The more lenient (schematic rather than axiomatic) reading of *EP-Independence* also fails for set theory. For example, typical presentations of ZF or ZFC also include an axiom positing the existence of an empty set, whose existence also follows from the Axiom of Infinity and simple operations. (Infinity posits the existence of a set containing the empty set and closed under the successor operation.) Indeed, the existence of an empty set follows from the existence of *any* set by the Separation Axiom. Modern set theory finds it useful to have a separate Empty Set Axiom, both for clarity and to investigate

[103] Kunen (2013: 134). [104] We are grateful to a referee for pressing this point.
[105] See, for example, Simpson (2009: 4) or Smith (2013: 210).

what set theory without the Axiom of Infinity might look like. Similarly, Replacement implies Separation (given the existence of the empty set) by a standard sort of argument. Separation is usually included partly for historical reasons but also to compare various axiomatisations of set theory with or without it. So even on this more lenient characterisation – that takes schemes as single principles – set theory does not respect *EP-Independence* even if it could easily be modified, without major consequences, to do so.

We note finally that the group-theoretic axioms contain no redundancy, so that group theory satisfies *EP-Independence*. In this respect, it is fairly typical of general theories.

To sum up, *EP-Independence* can be understood in a strict way as applying to axioms. Thus understood, group theory is the only one in our trio of theories that respects it; set theory and arithmetic do not. Understood more generously to cover schemes taken as single principles rather than their instances, *EP-Independence* is also respected by first-order arithmetic. It is not respected by set theory or some presentations of second-order arithmetic, but both could be modified to do so. What these cases suggest is that simplicity is an overriding virtue of an axiomatic presentation, and independence a secondary one. Which explains why, sometimes, a degree of axiomatic, and even schematic, independence is sacrificed for simplicity. Minor departures from *EP-Independence* in typical axiomatic presentations of arithmetic and set theory illustrate this point.

8.4 EP-Completeness

The fourth and last of the subsidiary principles requires theories to be complete, at least with respect to a certain specified class of truths.

We begin by considering general theories. Naturally, general theories are not required to be complete in the strong sense of negation-completeness. Group theory, for instance, should take in both abelian and non-abelian groups, so should not settle the claim that the group operation is commutative one way or the other.

That said, general theories are required to be complete in the narrow sense that they do not miss any facts expressible in the language of the theory that are common to every structure that the theory is trying to capture. But since stipulation is involved, completeness in this sense is easily secured, since we can just fix the meaning of, say, 'group', to cover exactly the structures satisfying the stipulated axioms of the subject we call 'group theory'. The facts common to all such structures expressible in the language of group theory are exactly the first-order logical consequences of the axioms of group theory. Since first-order logic is complete, these consequences are all derivable from the axioms. Perhaps the only way completeness in this weak sense could fail is if, say, the word 'group' were used to mean something like 'whatever *these* structures [pointing to some structures]' have in common but the axioms failed to accurately capture the structures' relevant commonalities. (For example, the structures might all be groups but the proposed axioms those for monoids.) This sort of mismatch between intended structure and proposed axiomatisation can occur early on in the development of a general theory but is unlikely to persist.

Moreover, the interest of the narrow reading is meagre, given how limited the facts about groups that we can express in first-order group theory are. On a broader reading of completeness, we might consider a known or knowable fact of group theory to be one that mathematicians (especially group theorists) would count as such. On this reading, even the weakest form of *EP-Completeness*, according to which the known truths of the subject follow from the axioms, would fail, since the known facts in the broader sense quite obviously include many claims not expressible in the language of first-order group theory. The claim that one structure is isomorphic to another, for example, is not expressible in this language, nor are generalisations over all finite groups, nor more generally are any claims involving the words 'finite' or 'infinite' (since these notions are not first-order expressible),[106] and so on. Similar considerations could be raised about other general theories. We conclude that general theories do not typically satisfy *EP-Completeness* in the broader, more interesting sense.

Let's turn now to specific theories, which are a different kettle of fish. As discussed in §2, we do not require that the consequence relation of a Euclidean theory be formal deduction; geometrical intuition or the drawing of diagrams, for instance, might be allowed to play an essential role in proofs and constructions. This makes claims of completeness very difficult to assess for unformalised theories, because whether some theorem has a proof can depend on what modes of inference we take to be intuitively admissible, and whether some problem has a solution can depend on the whether some diagrammatic construction is properly geometrical. This has been a rather fraught topic of discussion through the ages,[107] and one can readily imagine similar problems about the extent of intuition for areas other than geometry. We thus restrict our attention to specific theories with a formal deductive relation, as is the norm in contemporary mathematics. For them, it is natural to suppose that completeness in the sense of negation-completeness is an ideal. We consider first the strongest version of *EP-Completeness*, namely, that a theory of some particular mathematical structure-type should derive all facts about it.

[106] For example, the notion of being torsion-free is not first-orderisable; see Paseau and Leek (2022: 6). (A group is torsion-free if the identity is the only element with finite order.)

[107] Indeed, it is a central concern of Descartes in the *Geometry*.

Unfortunately, that ideal is unattainable. This follows from Gödel's First Incompleteness Theorem. Incorporating a strengthening owed to Rosser, we may state it as saying that any recursively axiomatised and consistent theory containing or interpreting the axioms of a first-order system known as Robinson Arithmetic is negation-incomplete. To unpack this: a theory is recursively axiomatised just when the properties of being a well-formed formula of its language, being an axiom, and being a correct proof are all recursively decidable. Robinson Arithmetic is a very weak theory of arithmetic, which any theory of arithmetic worth its salt will imply, and certainly any foundation of mathematics should. The word 'containing' in the statement of the theorem is crucial and gives the theorem its great strength; adding more axioms to the theory is no fix, without compromising the consistency or recursive axiomatisability of the theory. There are extremely weak theories which circumvent the Gödelian limitation, such as Tarski's axiomatisation of elementary geometry, which *is* deductively complete and moreover is recursively decidable. But Tarski's geometry is only a fragment of Euclidean geometry and does not treat the full theory.[108] Richer theories of geometry are incomplete because they can interpret Robinson Arithmetic (e.g. an infinite succession of points may be identified with the numbers). This means that the strongest ideal of completeness cannot be attained even in the original home of the EP, let alone further afield in the richer areas of contemporary mathematics.

The consistency of a theory follows from the soundness of the system, part of *EP-Truth*. And the inclusion of Robinson Arithmetic is a very weak requirement, as noted, which must be accepted whether arithmetic is axiomatised on its own, or is carried out in some other theory, be it set theory or geometry. Note that the theory's recursive axiomatisability follows from *EP-Finite*, assuming that the language and deductive system of the theory behave standardly. (Roughly put, we must be able to tell, of a formula that it is a formula, and of a proof that it is a proof.) These sorts of requirements have been (in a loose form) implicit throughout the history of the EP, at least until the twentieth century, and were not explicitly stated in §2 because they are taken for granted. Which is why, say, True Arithmetic, the theory consisting of all true arithmetical claims in the language of PA^1, cannot serve the EP's epistemological purposes.[109]

In sum, Gödel's First Incompleteness Theorem sinks the strongest version of *EP-Completeness* for specific theories. Of course, no one prior to the twentieth century, or indeed prior to 1930, could have anticipated this result, which goes some way towards explaining the EP's enduring appeal up to the twentieth century.

[108] Tarski and Givant (1999: 175). [109] See Wrigley (2022b, §6) for more on this point.

One reply to this line of thought is to opt for semantic, as opposed to deductive, completeness. To recap some familiar ideas, let's consider the case of arithmetic, which, as we saw, admits of a first-order axiomatisation PA^1 or a second-order axiomatisation PA^2.[110] Since PA^1 is a first-order theory, it follows from the soundness and completeness of first-order logic that the logical consequence relation (symbolised by '$\vDash$') is coextensive with the deduction relation (symbolised by '$\vdash$'). By Gödel's First Incompleteness Theorem, PA^1 is deductively incomplete, assuming it is consistent. This means that for any sound and complete proof system for first-order logic, there is a sentence ϕ in the language of PA^1 such that

$$PA^1 \nvdash \phi \text{ and } PA^1 \nvdash \neg\phi.$$

In contrast, all the models of PA^2 are isomorphic to one another: PA^2 is a categorical theory. It must therefore be semantically complete, because categoricity implies semantic completeness. That is to say, for all ϕ in the language of PA^2,

$$PA^2 \vDash \phi \text{ or } PA^2 \vDash \neg\phi$$

The friend of the EP may therefore counter that the Euclidean ideal should be semantic, rather than deductive, completeness ('$\vDash$' rather than '$\vdash$'). The logic in which theories are couched cannot then be first-order, the reason being that for first-order logic, semantic and deductive consequence coincide. But it might be second-order logic, or some other logic stronger than first-order.

Against this manoeuvre, three objections may be raised. The first is that it is unfaithful to the EP. Logicians' conception of logical consequence has varied over millennia,[111] with some stressing necessary truth-preservation, a more semantic notion, and others deducibility, a more syntactic notion. But when it comes to the EP, the relevant notion of consequence has always been deducibility, since the aim is to derive theorems so that we may come to know them.

A second, closely related, objection: this is as it should be. The First Incompleteness Theorem applies just as much to PA^2 equipped with a deductive system as it does to first-order theories such as PA^1. For any sound second-order deductive system, there is some ϕ in the language of PA^2 such that

$$PA^2 \nvdash \phi \text{ and } PA^2 \nvdash \neg\phi.$$

The set of PA^2's logical consequences therefore outstrips that of its theorems in any sound deductive system, showing that the EP's epistemological ends would

[110] It is particularly important to stress here that the second-order axiomatisation is in second-order logic with full semantics: the higher-order quantifiers range over *all* functions and relations over the domain.

[111] See chapter 1 of Griffiths and Paseau (2022).

not be served by this conception of completeness. More generally, the First Incompleteness Theorem rules out a deductively complete axiomatisation of any mathematical theory (that satisfies its conditions) in any logic.

The third objection is that even the move to a second-order axiomatisation based on semantic consequence will not overcome incompleteness in all areas. In particular, no second-order axiomatisation of standard set theory is complete. Take, for example, ZFC^2, the second-order version of ZFC set theory. As Zermelo (1930) showed, ZFC^2 is quasi-categorical – all its models are linearly ordered with respect to embedding – which gives it a partial kind of semantic completeness, but not the full version; the theory has many non-isomorphic models of differing sizes. Contemporary set theory, then, is an example of a specific mathematical theory that lacks a semantically complete axiomatisation. We conclude that the strong version of *EP-Completeness* neither does, nor should, in general apply to all axiomatisations.

As highlighted earlier, the completeness requirement is schematic. Gödel's First Incompleteness Theorem sinks its strongest form, but what about weaker versions?

Unfortunately, these are no more plausible. To see this, consider the weakest version of the requirement, which claims only that the *known* truths of the relevant domain are all deducible from the axioms. Gödel's theorems do not just imply that recursively axiomatised theories containing Robinson Arithmetic are deductively incomplete if they are consistent. The standard method of proving these theorems also provides examples of undecidable statements. In particular, a canonical consistency sentence of a theory meeting the requirements for Gödel's theorem is undecided by the theory (this is Gödel's Second Incompleteness Theorem). Such a sentence expresses (via Gödel numbering) the proposition that the theory is consistent, and this causes problems even for weak versions of *EP-Completeness*, as we now explain.

According to the Euclidean, the axioms of a suitable theory are all known to be true, and deduction from them is truth-preserving. By induction, all theorems of the theory are true also. Since any true set of sentences is consistent, it follows from a minimal amount of meta-theoretic reflection that the theory is itself consistent. Supposing, as the Euclidean must, that the results of deduction from known premises are known, it follows that the theory is known to be consistent. Since the consistency sentence of a theory is a sentence in the language of that theory, it also expresses a standard claim about the objects of the theory. For example, the consistency sentence of Peano Arithmetic will say that some number (coding the claim that $0 = 1$) does not satisfy a particular numerical predicate (coding deducibility from the axioms). Since that claim is known to be logically equivalent to a known claim (namely, the consistency of the theory), it is once again known by the Euclidean theorist. Hence there is some known claim about the domain which does not follow deductively from the axioms. Moreover, since what is known is knowable, there is some *knowable* claim about the objects of the domain which does not follow from the axioms of the theory, scuppering an intermediate form of *EP-Completeness* as well.

A caveat: the argument we advocate is not always available; in particular, it is not available for a (consistent, sufficiently strong, recursively axiomatised) theory with a truth predicate that applies to any sentence of the language (including sentences containing the truth predicate). In such a theory, we would be able to formulate the above argument from the truth of the axioms to the consistency of the theory, but the second incompleteness theorem shows that this must go wrong somehow.[112] Exactly how it goes wrong is sensitive to the behaviour of the truth predicate.[113] But the kind of theories that we're considering, such as arithmetic and set theory, do not standardly include a truth predicate at all, and for these kinds of theory, it is unproblematic to infer the theory's consistency from the truth of its axioms if the deductive rules are truth-preserving. For our purposes, this is sufficient, since theories including this sort of predicate are not mainstream fare in mathematics today, nor historically the EP's concern. They offer an interesting exception to our Gödelian argument, but an exception no more troubling for our assessment of the EP than the existence of non-recursive theories, or theories so weak as to not contain Robinson Arithmetic. We conclude that Gödel's second theorem sinks the weaker versions of *EP-Completeness* just as thoroughly as the first theorem sinks their strong counterpart.

In the last two sections, we compared mathematical practice and the EP, focusing on a trio of theories: set theory, arithmetic, and group theory. We summarise the discussion by means of Table 1. Table 1 ignores the complexity of the discussion in §§7–8, so is no substitute for it; but it may nevertheless be helpful as a summary of the headline findings. (The appropriate versions of *EP-Completeness* are as discussed in §8.4.).

9 What Should Replace the EP?

In §§7–8, we argued that the EP applies to contemporary mathematics in some ways but not others. If the EP is at best a partially accurate picture of mathematical epistemology, what should replace it? This question is too large for us to answer here, but we make a start on it by exploring some features of

[112] Even for such theories, the consistency of the theory might be knowable, or indeed known, by other means. One of us has argued elsewhere that we can have non-deductive, non-testimonial knowledge of some mathematical claims. See Paseau (2015) and (2023), though the examples there were not consistency statements.

[113] See Field (2006) for more on this. Thanks to an anonymous referee for highlighting this point.

Table 1 Summary

	Truth	Self-Evidence	Flow	Finite	General	Independence	Completeness
Set Theory	Yes	No (not all)	No	No (axiomatic) Yes (schematic)	Yes (allowing defs)	No (axiomatic) No (schematic)	No
Arithmetic	Yes	No (not all)	No	No (PA^1, axiomatic) Yes (PA^2, axiomatic, and PA^1, schematic)	Yes (allowing defs)	No (PA^1, axiomatic, PA^2 some versions) Yes (PA^1 schematic, PA^2 some versions)	No
Group Theory	Yes (by stipulation)	Yes	No	Yes	Yes	Yes	Yes (narrow) No (broad)

contemporary mathematical epistemology, all of them important parts of post-Euclidean thinking. Our focus will be on set theory, for reasons that will emerge. Together with a few surviving insights from the Euclidean Programme, they yield a messy but rich picture of mathematical knowledge that points towards a worthy successor to the EP.

9.1 Intrinsic vs Self-Evident

Recall the distinction drawn in §7 between intrinsic and extrinsic justification for an axiom.[114] Our first suggestion is that, as far as the epistemology of mathematics is concerned, it is more useful to think in terms of intrinsically justified axioms than self-evident ones.

Self-evident axioms are intrinsically justified. Definitional or quasi-definitional axioms, such as set theory's Axiom of Extensionality, are excellent candidates for self-evidence, as is the arithmetical axiom that 0 is no number's successor. But not all axioms enjoy this kind of status. Consider, for instance, set theory's Axiom of Infinity. This axiom does not enjoy the intuitive force of even the most debatable arithmetical axiom, the induction principle (see §7.2). Denying it does not seem contradictory, nor even conceptually challenging in the way that denying the Parallel Postulate, which is not straightforwardly true, might at first sight appear.

Yet, it would be too quick to rule that *any* justification that the axiom enjoys is thereby extrinsic. For intrinsic justification may also accrue to an axiom if it follows from a conception of the relevant subject area. Gödel, characterised by Lakatos as a *new* Euclidean (1976: 217), pioneered this conception of intrinsic justification, by discussing set-theoretic axioms implied by the concept of set (1964: 260).[115] Where the EP speaks of self-evident principles *tout court*, post-Gödel we tend to think of principles that are obvious/evident *on a conception*.

To be evident on a conception is to be intrinsically justified, albeit in a relative sense. The Axiom of Infinity, while it also enjoys a degree of extrinsic justification, is an obvious and fundamental truth about the universe of sets on the iterative conception. Gödel contrasts this with the naïve conception and describes it as the concept of set 'according to which a set is something obtainable from the integers (or some other well-defined objects) by iterated application of the operation "set of", not something obtained by dividing the totality of all existing things into two categories' (1964: 259). As Boolos highlights, a basic feature of the iterative conception is that the universe of

[114] Some complain that the intrinsic/extrinsic distinction is vague and should be understood as involving different facets of explanatoriness. See Barton, Ternullo, and Venturi (2020). We take it for granted, however, that the distinction is workably clear.

[115] For more on Gödel's conception of intrinsic justification, see Wrigley (2022a).

sets is formed in stages, and it is an equally central feature of the iterative conception that this sequence goes on to at least ω, the first infinite ordinal, and beyond (1971: 220–2). So the Axiom of Infinity is not just extrinsically justified: we have a tolerably clear conception of the universe of sets in which the axiom evidently holds, conferring upon it a degree of intrinsic justification as well.

Of course, questions remain about the extent to which various axioms are intrinsically justified, for example whether the Axiom of Choice, which states that any non-empty set of disjoint non-empty sets has a choice set, is obvious on the iterative conception.[116] We do not wish to weigh in on these considerations here; for our purposes it is sufficient to underline that the EP is blind to the important distinction between axioms which are *self*-evident and those which are evident *on a conception.*[117]

One might object on the Euclidean's behalf: wasn't relativity to a conception always there? Consider, for example, the induction principle in arithmetic. It is senseless to say that induction is evident completely independently of a conception of the natural numbers. And no self-respecting Euclidean would have ever said as much. Rather, to say that the induction principle is self-evident means that one needs no proof of the principle once one has grasped the axiom and the intended conception of the natural numbers lying behind it.

In response, we observe that the notion of self-evidence is not one that comes in degrees, unlike evidence on a conception. As we saw above, even the axioms of arithmetic, which we do not take to require a proof, are not all alike in the degree of intrinsic evidence which they enjoy. Nor does every coherent conception of the natural numbers justify the standard forms of the induction principle. Both points tell against the *self*-evidence of the induction principle in the standard sense that adherents of the EP have insisted upon.

We saw in §§7–8 that part of the Euclidean Programme has survived to the modern day; *EP-Truth*, *EP-Finite*, and *EP-General* are all cogent requirements to place on mathematical theories. We might then add to these a replacement for *EP-Self-Evidence*: axioms can be justified because they are evident *relative to* a conception of the subject matter of the theory. The strength of that justification will depend on the forcefulness of the underlying conception. But this kind of

[116] Hauser (2005) addresses the question of whether the Axiom of Choice is self-evident. Another important question is how many conceptions underlie a particular subject area. In the case of set theory, is the iterative conception the only underlying conception, or are there at least 'two thoughts behind set theory', as George Boolos put it, alluding to the iterative *and* Limitation of Size conceptions? See Boolos (1971) and (1989), both critically assessed in Paseau (2007).

[117] A terminological point: anyone who disagrees with our choice of 'intrinsic' for what we mean by it should feel free to replace it with another word. Our substantive point is that an important epistemological category is missed if we focus only on self-evident axioms.

relative evidence is not the end of the story; as we shall now observe, extrinsic evidence also has a major role to play in modern mathematical epistemology.

9.2 Extrinsic Justification

The EP has several blind spots. One is that it ignores non-deductive justification in mathematics (see Paseau 2023 for more on this sort of justification). But perhaps the biggest blind spot in the EP, from a contemporary perspective, is that it ignores the justification of an axiom through its consequences. Extrinsic considerations play an absolutely central role in the modern epistemology of mathematical theories. Extrinsic justification for an axiom is analogous to the kind of justification enjoyed by well-established scientific principles. In this section, we shall confine our attention to arithmetic and set theory, mostly leaving group theory behind. Considerations of the fruitfulness of the consequences of the group axioms relate less to their justification per se, since they are self-evidently true by stipulation, and more to the motivation for formulating and developing the theory in the first place.

The first kind of extrinsic justification to be discussed we call *strictly regressive justification*. It is analogous to the kind of justification enjoyed by a scientific postulate which contributes directly to the empirical adequacy of a theory; that is, a postulate without which a range of data could not be accounted for. Lakatos, as we saw in §2, takes the Empiricist Programme to be bottom-up, in contrast to the Euclidean Programme's top-down approach. Strictly regressive justification plays a leading role in the Empiricist Programme. A doctrine along these lines is most clearly articulated by Russell, who draws a 'close analogy between the methods of pure mathematics and the methods of the sciences of observation' (1907: 572). It is common enough to conceive of natural-scientific propositions as being divided into two broad kinds (whether or not we take those kinds to be disjoint or sharply delimited): the data and the general principles or laws. On this conception, data are empirical propositions that we take to be facts, and general principles those propositions formulated in order to predict the facts, in concert with auxiliary hypotheses. Indeed, the prediction of the data is the primary means of verifying these principles or laws, either individually or collectively. The data have their own special epistemological status, but whether or not the general principles are intrinsically plausible, we take them to be true if they predict all the data and don't predict anything we know to be false.

By analogy, mathematical axioms are supposed to be verified by 'predicting' (i.e. deductively implying) propositions of some privileged kind identified as the data. This is broadly Russell's view of the matter, which he calls the 'regressive method' of justifying axioms. It plays a major role in his own foundational system; for example, in *Principia Mathematica*, he (with Whitehead) had this to say about the controversial Axiom of Reducibility:

> That the axiom of reducibility is self-evident is a proposition which can hardly be maintained. But in fact self-evidence is never more than a part of the reason for accepting an axiom, and is never indispensable. The reason for accepting an axiom, as for accepting any other proposition, is always largely inductive, namely that many propositions which are nearly indubitable can be deduced from it, and that no equally plausible way is known by which these propositions could be true if the axiom were false, and nothing which is probably false can be deduced from it. (Whitehead and Russell 1910/1927: 59)

It is essential to the Empiricist Programme that certain propositions are identified as being data and that these have a special epistemological status which explains their role in the programme. Various manifestations of the Empiricist Programme will have different ideas about which propositions are properly classified as data, or which propositions are lit by the 'natural light of Experience', as Lakatos puts it.

Russell himself is not particularly informative on what to count as data, nor does he give a detailed explanation of their privileged epistemic status. But he singles out propositions of a simple kind, which appear to be justified by observation of large numbers of correct instances, and no observation of false instances. Take, for example, the proposition that $2 + 2 = 4$, which ought to count as common fact if anything does. Russell conjectures that the historical sources of this belief (or its 'empirical premises', as he calls them) will be various observations from everyday life, such as ancient shepherds repeatedly noticing that two pairs of sheep are always four sheep. Outside the epistemology seminar, serious doubt cannot be entertained about such propositions. The truths of elementary arithmetic are therefore far more evident than the axioms of any system from which they could be derived. This leads Russell to conclude that the method of discovering and justifying foundational principles in mathematics is 'substantially the same as the method of discovering general laws in any other science' (1907: 573). Given the similarity of methods of justification, it is unsurprising that for Russell the degree of verification obtained by axioms in mathematics is similar to the degree which may be claimed for the laws of physics. As he puts it, 'when the general laws are neither themselves obvious, nor demonstrably the only possible hypotheses to account for the [data], then the general laws remain merely probable' (1907: 573).

But that does not mean that strictly regressive justification is not epistemologically significant. Obvious elementary statements, particularly those drawn from arithmetic, place a severe constraint on which axiomatic theories we can regard as true, even if all such statements are readily provable in a fairly weak theory; for example, we should have to reject as false a theory which implied that $2 + 2 \neq 4$. Data also constrain our axiomatic theories in another way; contradicting the data falsifies a theory, but merely failing to imply the data shows a theory to be inadequate. For example, Robinson Arithmetic, lacking the axiom scheme of induction, cannot prove that $x+y=y+x$; so, it is not a serious candidate for axiomatising arithmetic.

It is important to stress that these observations do not *contradict* the EP, strictly speaking. We did not bake a chauvinism principle into the EP (see §5.1), so perhaps a Euclidean could provide a supplement to their theory which accounts for our knowledge of elementary arithmetical propositions and whatever else we might want to count as data. It would then be open to them to explain the above phenomena. Proving the negation of a datum from some axioms would show them to be false, thereby failing on the count of *EP-Truth*; and showing that the axioms fail to derive some data would show that they fell too far short of *EP-Completeness*.

The problem is rather that it is as close to a plain fact as can be found in mathematical epistemology that certain non-axioms are known without proof. And the EP simply has nothing illuminating to say about this fact. Elementary arithmetical propositions, in particular, do not fit the epistemological picture painted by the core principles of the EP. Even somebody who thought that the axioms of PA are self-evident would hardly be *more* confident that $2 + 2 = 4$ upon learning that the proposition is provable in PA. Rather, they should become more confident that PA correctly axiomatises our conception of the natural numbers. So, *EP-Flow* is silent on the epistemology of the data, and *EP-Self-Evidence* ignores an important part of the epistemology of axioms.

Just as we identified a modern replacement of *EP-Self-Evidence* in §9.1, in §9.2 we have found a modern replacement for *EP-Flow*. Axioms can transfer their epistemic good standing to theorems via proof, to at least some degree; but additionally, the proof of a mathematical datum may transfer the epistemic good standing of that datum to the axiom system in which the proof is given. Providing an epistemology of axioms in terms of theorems which can be inferred from them, including giving an appropriate delineation of the data and explanation of their epistemic status, is a pressing task that we can add to the agenda for the philosophy of mathematics today.

Extrinsic justification of a second kind is also possible. This sort of justification is analogous to the kind enjoyed by scientific postulates which enhance the simplicity, economy, unifying power, and so on, of the theories to which they belong. Extrinsic justification of this kind, in terms of *theoretical virtues*, is, in mathematics, confined largely to foundational areas such as set theory. Here, the axioms do not enjoy the same degree of intrinsic justification as in other areas, such as arithmetic, and not all of the axioms enjoy much strictly regressive support either. This applies to both widely accepted axioms, such as Replacement and Choice,[118] but also to more speculative principles, such as large cardinal axioms. Although set theory without Replacement or Choice is fairly weak by the standards of a foundational theory, it is still very powerful by the general standards of mathematics, and any elementary proposition that can plausibly be regarded as data is likely to be derivable already in this theory.

It is in accounting for axioms like these, which go beyond the bare foundational necessities, that justification in terms of theoretical virtue is most prominent today. Once again, the modern conception of this kind of justification in mathematics is found in the work of Kurt Gödel. As he puts it,

> ... even disregarding the intrinsic necessity of some new axiom, and even in case it had no intrinsic necessity at all, a probable decision about its truth is possible also in another way, namely, inductively by studying its 'success'. Success here means fruitfulness in consequences, in particular in 'verifiable' consequences, i.e., consequences demonstrable without the new axiom, whose proofs with the help of the new axiom, however, are considerably simpler and easier to discover, and make it possible to contract into one proof many different proofs. The axioms for the system of real numbers, rejected by the intuitionists, have in this sense been verified to some extent, owing to the fact that analytical number theory frequently allows one to prove number-theoretical theorems which, in a more cumbersome way, can subsequently be verified by elementary methods. A much higher degree of verification than that, however, is conceivable. There might exist axioms so abundant in their verifiable consequences, shedding so much light upon a whole field, and yielding such powerful methods for solving problems (and even solving them constructively, as far as that is possible) that, no matter whether or not they are intrinsically necessary, they would have to be accepted at least in the same sense as any well-established physical theory. (1964: 261)

This idea has been hugely influential. The role of theoretical virtues in modern mathematics, such as a theory's simplicity or its ability to solve open problems, speed up proofs, and unify disparate areas are by now well-acknowledged. The contemporary philosopher of mathematics who has done most to bring the relevant ideas to the fore is Penelope Maddy. As she has emphasised, many of the extrinsic reasons for believing the axioms of ZFC and some of its extensions are much richer

[118] See Potter (2004: §§14.6 and A.3) for a more detailed discussion of why the regressive case for these axioms is unpromising.

than the simple deducibility of data described by Russell. These reasons are catalogued in Maddy (1988, 2011) and other publications.

Thanks to the work of Gödel, Maddy, and others, something like a consensus seems to have been reached about how foundational mathematical theories are justified and what a good axiomatisation looks like, at least in outline form. This does not mean that there is no disagreement over details, still less that no arguments exist about which axiom candidates meet the criteria. But the broad shape of a justificatory project is generally agreed upon. That shape does not, even in outline, match that of the EP, which has just as little to say about the value of theoretical virtues as it does about regressive justification.

To sum up §9, we think that the messy and complicated interaction of intrinsic and extrinsic considerations which we see exhibited in current mathematical practice is rightly described as the successor to the EP, in set theory at least. In principle (usually in practice too), the justification for an axiom is both intrinsic and extrinsic, the exact proportions varying depending on the axiom.

We have, of necessity, touched on these issues all too briefly, because the star of the show, in this essay, is the EP and not its contemporary replacement. What we hope to have made clear in §9 is that the contemporary debate in the area in which the EP is most deficient – set theory, or equivalently any foundation of mathematics – takes a very different shape to the one imagined by Euclideans. And even the epistemology of more elementary areas, such as arithmetic, is more complicated than the advocates of the EP appreciated.

10 Summary

In §§1–2, we introduced and reconstructed the Euclidean Programme. In §§3–6, we compared this reconstruction with some historical accounts. In §§7–8, we assessed the EP against contemporary practice, focusing mainly on three areas of mathematics: set theory, arithmetic, and group theory. We showed that, despite its having held sway for millennia, the EP is no longer tenable as a general and fully accurate account of all mathematics. Some Euclidean tenets are in place today in many areas of mathematics, but not all of them are in place throughout. In §9, we sketched in a very preliminary way how contemporary epistemology of mathematics goes beyond the EP.

Einstein, in the 1934 article quoted in the introduction, also wrote that '[c]onclusions obtained by purely rational processes are, so far as Reality is concerned, entirely empty' (1934: 164). Einstein had physics in mind here. But if 'purely rational processes' means the Euclidean method conceived along the lines of the EP, then his point has some validity as far as contemporary mathematics goes.

References

C. Adam and P. Tannery (eds.) (1897–1910), *Oeuvres de Descartes*, vols. 1–12, Leopold Cerf.

Aristotle (4th century BC), *The Categories, On Interpretation, Prior Analytics*, transl. by H. P. Cooke and H. Tredennick, 1938 ed., Harvard University Press: Loeb Classical Library.

Aristotle (4th century BC), *Physics*, transl. by R. Waterfield, 2008 ed., Oxford University Press.

Aristotle (4th century BC), *Posterior Analytics, Topica*, transl. by H. Tredennick and E. S. Forster, 1960 ed., Harvard University Press: Loeb Classical Library.

A. Arnauld and P. Nicole (1683/2019), *La Logique ou l'Art de Penser* (5th ed.), 2nd Revised ed. by P. Clair and F. Gibral, Librairie Philosophique J. Vrin.

J. Barnes (ed.) (1993), Aristotle's *Posterior Analytics* (2nd ed.), Oxford University Press.

J. Barnes (2005), 'What Is a Disjunction?', in D. Frede and B. Inwood (eds.), *Language and Learning*, Cambridge University Press, repr. in his *Logical Matters* (2012), Oxford University Press: 512–37.

N. Barton, C. Ternullo, and G. Venturi (2020), 'On Forms of Justification in Set Theory', *Australasian Journal of Logic* 17: 158–200.

G. Boolos (1971), 'The Iterative Conception of Set', *Journal of Philosophy* 68: 215–31.

G. Boolos (1989), 'Iteration Again', *Philosophical Topics* 17: 5–21.

H. Bos (1981), 'On the Representation of Curves in Descartes' Géométrie', *Archive for History of Exact Sciences* 24: 295–338.

H. Bos (2001), *Redefining Geometrical Exactness*, Springer.

D. Bronstein (2016), *Aristotle on Knowledge and Learning: The* Posterior Analytics, Oxford University Press.

M. F. Burnyeat (1981), 'Aristotle on Understanding Knowledge', in E. Berti (ed.), *Aristotle on Science: The Posterior Analytics*, Antenore: 97–139.

A. Clairaut (1741), *Éléments de Géométrie*, ed. by H. Regodt (1853), Jules Delalain.

J. Clarke-Doane (2013), 'What Is Absolute Undecidability?', *Noûs* 47: 467–81.

J. Corcoran (1974), 'Aristotle's Natural Deduction System', in J. Corcoran (ed.), *Ancient Logic and Its Modern Interpretations*, Reidel: 85–131.

P. Corkum (2016), 'Ontological Dependence and Grounding in Aristotle', *Oxford Handbooks Online in Philosophy*, https://doi.org//10.1093/oxfordhb/9780199935314.013.31.

E. Craig (1996), *The Mind of God and the Works of Man*, Oxford University Press.

P. Crivelli (2004), *Aristotle on Truth*, Cambridge University Press.

K. Davey (2021), 'On Euclid and the Genealogy of Proof', *Ergo* 8: 54–82.

N. C. Denyer (2022), 'Diagrams and Proof in Euclid, Aristotle, and Plato' (ms.).

R. Descartes (1637/2001), *Discourse on Method, Optics, Geometry, and Meteorology*, transl. by P. Olscamp (Revised ed.), Hackett.

R. Descartes (1984a/vol. 1) & (1984b/vol. 2), *The Philosophical Writings of Descartes*, transl. by J. Cottingham, R. Stoothoff, and D. Murdoch, Cambridge University Press.

W. Detel (2012), 'Aristotle's Logic and Theory of Science', in M. L. Gill and P. Pellegrin (eds.), *A Companion to Ancient Philosophy*, Wiley: 245–69.

M. Detlefsen (2014), 'Completeness and the Ends of Axiomatisation', in J. Kennedy (ed.), *Interpreting Gödel*, Cambridge University Press: 59–77.

M. Domski (2009), 'The Intelligibility of Motion and Construction: Descartes' Early Mathematics and Metaphysics, 1619–1637', *Studies in History and Philosophy of Science* 40: 119–30.

M. Dummett (1973), 'The Philosophical Basis of Intuitionistic Logic', in his *Truth and Other Enigmas* (1978), Harvard University Press: 215–47.

A. Einstein (1934), 'On the Method of Theoretical Physics', *Philosophy of Science* 1: 163–9.

S. Feferman (1998), *In the Light of Logic*, Oxford University Press.

S. Feferman (2000), 'Does Mathematics Need New Axioms?', *Bulletin of Symbolic Logic* 6: 401–13.

H. Field (2006), 'Truth and the Unprovability of Consistency', *Mind* 115: 567–605.

G. Frege (1884/1953), *The Foundations of Arithmetic* (2nd ed.), transl. by J. Austin, Northwestern University Press.

G. Freudenthal (1988), 'La philosophie de la géometrie d'al-Fârâbî: Son commentaire sur le début du Ier et le début du Ve livre des Éléments d'Euclide', *Jerusalem Studies in Arabic and Islam* 11: 104–219.

H. Friedman (1971), 'Higher Set Theory and Mathematical Practice', *Annals of Mathematical Logic* 2: 325–57.

J-L. Gardies (1982), 'L'interprétation d'Euclide chez Pascal et Arnauld', *Les études philosophiques* 2: 129–48.

J-L. Gardies (1984), *Pascal entre Eudoxe et Cantor*, Vrin.

K. Gödel (1931), 'On Formally Undecidable Propositions of *Principia Mathematica* and Related Systems I' transl. by J. van Heijenoort in Feferman et al (eds.) (1986), *Kurt Gödel: Collected Works* (vol. 1), Oxford University Press: 144–95.

K. Gödel (1964), 'What Is Cantor's Continuum Problem?', in Feferman et al (eds.) (1990), *Kurt Gödel: Collected Works* (vol. 2), Oxford University Press: 254–70.

O. Griffiths and A. C. Paseau (2022), *One True Logic*, Oxford University Press.

E. R. Grosholz (1991), *Cartesian Method and the Problem of Reduction*, Oxford University Press.

K. Hauser (2005), 'Is Choice Self-Evident?', *American Philosophical Quarterly* 42: 237–61.

T. Heath (ed.) (1925), *The Thirteen Books of the Elements* (2nd ed.), vol. 1: Books I–II, vol. 2: Books III–IX, vol. 3: Books X–XIII, Cambridge University Press, 1956 Dover ed.

A. Heyting (1980), *Axiomatic Projective Geometry* (2nd ed.), North-Holland.

D. Hilbert (1899a), *Grundlagen der Geometrie*, transl. by L. Unger from the 10th German ed. as *Foundations of Geometry* (1990, 2nd ed.), Open Court.

D. Hilbert (1899b), 'Hilbert to Frege 29.12.1899', in G. Gabriel (ed.) (1980), *Gottlob Frege: Philosophical and Mathematical Correspondence*, University of Chicago Press: 38–43.

D. Hilbert (1925), 'On the Infinite', transl. by S. Bauer-Mengelberg in J. van Heijenoort (ed.) (1967), *From Frege to Gödel: A Source Book in Mathematical Logic 1879–1931*, Harvard University Press: 367–92.

E. V. Huntington (1911), 'The Fundamental Propositions of Algebra', in J. W. A. Young (ed.) *Monographs on Topics of Modern Mathematics*, Longmans, Green and Co.: 149–207.

A. Jaffe (1997), 'Proof and the Evolution of Mathematics', *Synthese* 111: 133–46.

R. Jeshion (2001), 'Frege's Notions of Self-Evidence', *Mind* 110: 937–76.

D. Jesseph (2022), 'Berkeley and Mathematics', in S. Rickless (ed.), *The Oxford Handbook of Berkeley*, Oxford University Press: 300–25.

W. Kneale and M. Kneale (1962), *The Development of Logic*, Oxford University Press.

K. Kunen (2013), *Set Theory* (Revised ed.), College Publications.

I. Lakatos (1962), 'Infinite Regress and Foundations of Mathematics', in J. Worrall and G. Currie (eds.) (1978), *Mathematics, Science, and Epistemology*, Cambridge University Press: 3–23.

I. Lakatos (1976), 'A Renaissance of Empiricism in the Recent Philosophy of Mathematics', *British Journal for the Philosophy of Science* 27: 201–23.

H. D. P. Lee (1935), 'Geometrical Method and Aristotle's Account of First Principles', *The Classical Quarterly* 29: 113–24.

F. Lewis (1920), 'History of the Parallel Postulate', *American Mathematical Monthly* 27: 16–23.

G. E. R. Lloyd (2014), *The Ideals of Inquiry: An Ancient History*, Oxford University Press.

J. Locke (1689/2004), *An Essay Concerning Human Understanding*, ed. by R. Woolhouse, Penguin.

W. G. Lycan (2001), *Real Conditionals*, Oxford University Press.

P. Maddy (1988), 'Believing the Axioms', *Journal of Symbolic Logic* 53: 481–511, 736–64.

P. Maddy (2011), *Defending the Axioms*, Oxford University Press.

P. Mancosu and M. Mugnai (2023), *Syllogistic Logic and Mathematical Proof*, Oxford University Press.

V. McGee (1985), 'A Counterexample to Modus Ponens', *Journal of Philosophy* 82: 462–71.

J. S. Mill (1882), *A System of Logic, Ratiocinative and Inductive* (8th ed.), Harper & Brothers.

M. Moriarty (2020), *Pascal: Reasoning and Belief*, Oxford University Press.

B. Morison (2019), 'Theoretical Nous in the Posterior Analytics', *Manuscrito* 42: 1–43.

I. Mueller (1974), 'Greek Mathematics and Greek Logic', in J. Corcoran (ed.), *Ancient Logic and Its Modern Interpretations*, Reidel: 35–70.

E. Nelson (1986), *Predicative Arithmetic*, Princeton University Press.

B. Pascal (1655/1991), *De l'esprit géométrique*, in *Pascal: Œuvres Complètes, vol. III: Œuvres Diverses (1654–1657)*, J. Mesnard (ed.): 360–428 (including commentary by Mesnard), Desclée de Brouwer.

M. Pasch (1882), *Vorlesungen über neuere Geometrie*, Teubner.

A. C. Paseau (2007), 'Boolos on the Justification of Set Theory', *Philosophia Mathematica* 15: 30–53.

A. C. Paseau (2011), 'Mathematical Instrumentalism, Gödel's Theorem and Inductive Evidence', *Studies in History and Philosophy of Science Part A* 42: 140–9.

A. C. Paseau (2015), 'Knowledge of Mathematics without Proof', *British Journal for the Philosophy of Science* 66: 775–99.

A. C. Paseau (2016), 'What's the Point of Complete Rigour?', *Mind* 125: 177–207.

A. C. Paseau (2023), 'Non-Deductive Justification in Mathematics', in B. Sriraman (ed.), *Handbook of the History and Philosophy of Mathematical Practice*, Springer, https://doi.org/10.1007/978-3-030-19071-2_116-1.

A. C. Paseau and R. Leek (2022), 'The Compactness Theorem', *Internet Encyclopedia of Philosophy*, https://iep.utm.edu/compactness-theorem.

R. Pasnau (2017), *After Certainty*, Oxford University Press.

M. Potter (2004), *Set Theory and Its Philosophy*, Oxford University Press.

Proclus (5th century AD/1970), *A Commentary on the First Book of Euclid's Elements*, transl. by G. R. Morrow, Princeton University Press.

W. D. Ross (1957), *Aristotle's Prior and Posterior Analytics* (Revised ed.), Oxford University Press.

B. Russell (1907), 'The Regressive Method of Discovering the Premises of Mathematics', in G. Moore (ed.) (2014), *The Collected Papers of Bertrand Russell*, vol. 5, Routledge: 571–80.

J. Schechter (2013), 'Rational Self-Doubt and the Failure of Closure', *Philosophical Studies* 163: 429–52.

S. Shapiro (1997), *Philosophy of Mathematics: Structure and Ontology*, Oxford University Press.

S. Shapiro (2009), 'We Hold These Truths to Be Self-evident: But What Do We Mean by That?', *Review of Symbolic Logic* 2: 175–207.

S. Simpson (2009), *Subsystems of Second Order Arithmetic* (2nd ed.), Cambridge University Press.

T. J. Smiley (1973), 'What Is a Syllogism?', *Journal of Philosophical Logic* 2: 136–54.

P. Smith (2013), *An Introduction to Gödel's Theorems* (2nd ed.), Cambridge University Press.

T. Sorell (2016), 'Knowledge (*Scientia*)', in L. Nolan (ed.), *The Cambridge Descartes Lexicon*, Cambridge University Press: 423–8.

A. Tarski (1994), *Introduction to Logic and to the Methodology of Deductive Sciences* (4th ed.), Oxford University Press.

A. Tarski and S. Givant (1999), 'Tarksi's System of Geometry', *Bulletin of Symbolic Logic* 5: 175–214.

B. Wardhaugh (2020), *The Book of Wonders: The Many Lives of Euclid's* Elements, William Collins.

A. N. Whitehead and B. Russell (1910/1927), *Principia Mathematica to *56* (2nd ed.), Cambridge University Press.

W. Wrigley (2022a), 'Gödelian Platonism and Mathematical Intuition', *European Journal of Philosophy* 30: 578–600.

W. Wrigley (2022b), 'Gödel's Disjunctive Argument', *Philosophia Mathematica* 30: 306–42.

E. Zermelo (1930), 'On Boundary Numbers and Domains of Sets: New Investigations in the Foundations of Set Theory', transl. by M. Hallett in W. Ewald (1996), *From Kant to Hilbert* vol. 2, Oxford University Press: 1219–33.

B. Zuppolini (2020), 'Comprehension, Demonstration, and Accuracy in Aristotle', *Journal of the History of Philosophy* 58: 29–48.

Acknowledgements

We are grateful to Penelope Rush and Stewart Shapiro for welcoming this Element into their series. We would like to thank Ben Morison, Janine Gühler, and Nick Denyer for comments on the Aristotle and Euclid sections, Sébastien Maronne and Doug Jesseph for their comments on the seventeenth-century material, and Benjamin Wardhaugh, Fabian Pregel, Mark Rothery, Owen Griffiths, Penelope Maddy, Stewart Shapiro, and Zachary Stanley as well as two anonymous CUP referees for comments on the entire manuscript. Thanks also to audiences at the Munich Center for Mathematical Philosophy in June 2023 and at the LSE's *Imre Lakatos Centenary Conference* in November 2022 for helpful comments and suggestions, and to Djebbar Ahmed, Hassan Amini, and Lennart Breggren for references to works on Euclid's *Elements* by Islamic mathematicians. Alex dedicates this book to Athena and (his) Penelope, and Wesley dedicates it to Chloë.

Cambridge Elements Ξ

The Philosophy of Mathematics

Penelope Rush

University of Tasmania

From the time Penny Rush completed her thesis in the philosophy of mathematics (2005), she has worked continuously on themes around the realism/anti-realism divide and the nature of mathematics. Her edited collection *The Metaphysics of Logic* (Cambridge University Press, 2014), and forthcoming essay 'Metaphysical Optimism' (*Philosophy Supplement*), highlight a particular interest in the idea of reality itself and curiosity and respect as important philosophical methodologies.

Stewart Shapiro

The Ohio State University

Stewart Shapiro is the O'Donnell Professor of Philosophy at The Ohio State University, a Distinguished Visiting Professor at the University of Connecticut, and a Professorial Fellow at the University of Oslo. His major works include *Foundations without Foundationalism* (1991), *Philosophy of Mathematics: Structure and Ontology* (1997), *Vagueness in Context* (2006), and *Varieties of Logic* (2014). He has taught courses in logic, philosophy of mathematics, metaphysics, epistemology, philosophy of religion, Jewish philosophy, social and political philosophy, and medical ethics.

About the Series

This Cambridge Elements series provides an extensive overview of the philosophy of mathematics in its many and varied forms. Distinguished authors will provide an up-to-date summary of the results of current research in their fields and give their own take on what they believe are the most significant debates influencing research, drawing original conclusions.

Cambridge Elements

The Philosophy of Mathematics

Elements in the Series

A full series listing is available at: www.cambridge.org/EPM

www.ingramcontent.com/pod-product-compliance
Ingram Content Group UK Ltd.
Pitfield, Milton Keynes, MK11 3LW, UK
UKHW022003190125
453752UK00007B/69

9 781009 494403